AMERICAN OASIS

AMERICAN OASIS

||||||||||||||||||||||||||||||||||||||||||||||||||||||||||||||||||||||||||||||||||||||||||||||||

FINDING THE FUTURE IN
THE CITIES OF THE SOUTHWEST

||||||||||||||||||||||||||||||||||||||||||||||||||||||||||||||||||||||||||||||||||||||||||||||||||||||||||||||

KYLE PAOLETTA

PANTHEON BOOKS   NEW YORK

Published in the United States by Pantheon Books,
a division of Penguin Random House LLC, New York,
and distributed in Canada by Penguin Random
House Canada Limited, Toronto.

Pantheon Books and colophon are registered
trademarks of Penguin Random House LLC.

Poems by Rosa Alcalá, Jimmy Santiago Baca,
Chase McCurdy, and Ophelia Zepeda reproduced with the
permission of the authors.

Portions of this book previously appeared in slightly different
form in *The Baffler, The Believer, Columbia Journalism Review,
High Country News, The Nation, The New Republic,* and *New York.*

ISBN: 9780553387377
LCCN: 2024946878

www.pantheonbooks.com

Jacket illustration by Sam Chivers
Jacket design by Linda Huang

Printed in the United States of America
First Edition

9 8 7 6 5 4 3 2 1

to Mama and Bapee
for making me New Mexican

Habiéndose de edificar en la ribera de qualquier río de la parte del oriente de manera que en saliendo el sol dé primero en el pueblo que en el agua.

In case of building on the bank of some river, they shall arrange the settlement in such a way that the sun may shine on the town before it shines on the water.

—*Transcripción de las ordenanzas de descubrimientos, nueva población y pacificación de las Indias* (1573)

I am crying from thirst.
I am singing for rain.
I am dancing for rain.
The sky begins to weep,
for it sees me
singing and dancing
on the dry, cracked
earth.

—Alonzo Lopez (1969)

CONTENTS

Map xi

Preface xiii

PART I

THEY ARE MANY AND KEEP COMING

MARROW 3

DESTINY 43

RECALCITRANCE 69

CHIMERA 97

PART II

EL PUEBLO UNIDO
JAMÁS SERÁ VENCIDO

DEFIANCE 121

COMBUSTION 159

PART III

OUT OF MANY, UNO

VIABILITY 181

MUTUALISM 201

SPRAWL 235

AFTERWORD 255

ACKNOWLEDGMENTS 261

NOTES 265

BIBLIOGRAPHY 301

INDEX 309

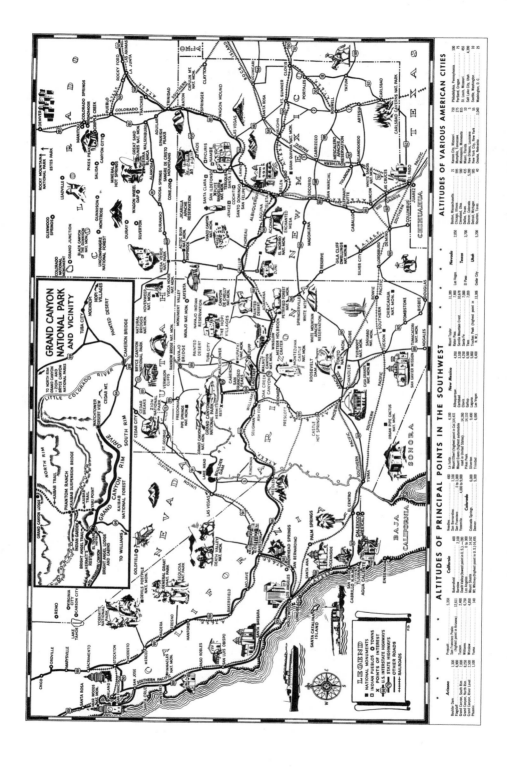

Preface

Heading east from the Pacific Coast on I-8, the border of the desert Southwest is obvious. Climb the steep, pine-studded Laguna Mountains in eastern San Diego County and the landscape suddenly gives way. Pick an overlook, and the sculptural badlands of Anza-Borrego unfurl before you, meandering up into a ridge that soon disappears into another dry valley flanked by more mountains, the basin and range pattern repeating into the soft infinity of the horizon. Or take the funicular up to the leafy, ten-thousand-foot summit of Mount San Jacinto above Palm Springs to see how the pinkish Little San Bernardino Mountains tower over the resort city's grid of mid-mod subdivisions and golf courses, several layers of silhouetted transverse ranges peeking through the gaps in their limpid wall of stone.

Head for the desert country from the other direction, and the geographic distinctions are more vague. Following I-10, the land turns arid in the barrens of West Texas, which are separated from the rest of the state by a hundred-mile swath of sun-bleached scar tissue, the earth littered with oil wells that form a vast, chaotic web spun by invisible geology, not surface terrain. Only once the brown earth crinkles up into the raw, prehistoric coral reef we call the Guadalupe Mountains can you be sure you've entered the great Chihuahuan Desert, which covers as much territory as the nation of France and extends in a parched, mountainous ribbon all the way from San Luis Potosí in the center of Mexico to the southern fringes of Albuquerque.

The Chihuahuan touches the Sonoran Desert near the border of Arizona and New Mexico, which, in turn, butts up against the Mojave in Southern California. Together, these deserts host the Southwest's largest cities and the overwhelming majority of its population—to their north, the highlands of the Colorado Plateau and the salt flats of the Great Basin represent one of the world's most spectacular arid landscapes, but harbor few of the necessary ingredients for creating a city.

Each desert is home to a signature cactus: the Mojave has the Joshua tree; the Sonoran, the saguaro. In the Chihuahuan, it's the somewhat less charismatic lechuguilla, a squat agave that arrests the attention of the visitor only after a dozen years of growth, at which point it has stored enough energy to shoot a stalk ten feet into the air that briefly flowers into a fiery column of red and yellow before the entire plant dies.

The O'odham peoples of the Sonoran made syrup from saguaro fruit and protein-rich meal from its seeds, while the cactus's long ribs were used as a lightweight building material for shade structures. In the Mojave, the Cahuilla wove the fibers of the Joshua tree's spines into baskets and clothing. Up and down the length of the Chihuahua, everyone from the Tarahumara to the Jumano made bow strings from the lechuguilla's leaves, produced soap from its roots, and fermented its sap into pulque, tequila, and mescal. Despite its popular connotation of nothingness, the desert provides.

Still, there's a difference between human subsistence and long-tenured community. For a village to grow to a city of a hundred thousand people, then a million, then five—well, not just any stretch of arid landscape will do. A metropolis can only sprout in the desert if there is water nearby, ideally in the form of an ever-replenishing river. What about topography? Surely there is something about this particular stretch of the Rio Grande or the Salt River that attracted settlers to it over any other? Say there are mountains, then. Big ones, sizable enough to delimit a shift from one landscape to another. Now let's give those mountains a name: Frenchman, Franklin, Camelback. Santa Catalina. Sandia.

The first misapprehension about the great cities of the Southwest is that they appeared out of the thin, high desert air; that their placement was as arbitrary as the logic governing their expansion. *Why build there?* The answer can be found in eons of churning stone. The earth created conditions such that when humans arrived at this coincidence of elevation and water, they could find nowhere better to go.

Whatever connotation of nothingness the word *desert* may harbor, people have been living here for millennia. Indeed, the modern American Southwest is merely the latest iteration of a pattern of human habitation that extends back to the first people who arrived on this continent from Asia. Only in the colonial era did the idea of these deserts as fundamentally inhospitable take hold. While the Spanish originally came searching for cities of gold and Anglos eventually followed with hopes of likewise extracting whatever wealth they could from the land, both sets of settlers saw the yawning, xerothermic province of thorny plains and ruddy mesas that extends from the Rockies to the Sierra Nevada as a vacancy to be filled. That Tewa, Diné, and Manso peoples were already present when the first Europeans arrived was incidental to the project of empire. The same goes for the genealogy and tradition that had rooted those Indigenous peoples in their homelands. The ancestors of the O'odham are known by archeologists as the Hohokam, whose canal-fed farms once supplied tens of thousands of people in the Salt River Valley—at its height, their network of villages may have composed the largest settlement in North America. Centuries earlier, the precursors of the Puebloan peoples of the Rio Grande built stately cliff dwellings at Mesa Verde, Chaco Canyon, and the Pajarito Plateau.

Never mind all that. For the desert to be mastered by those of European stock, raised to believe that a place needed to be green in order to give life, the region had to be understood as a vacancy in need of filling. Álvar Núñez Cabeza de Vaca was the first Spaniard to chronicle a journey into what's now the Southwest. He survived a disastrous expedition that began in Florida in 1528 but morphed into an arduous overland trek to Mexico City after most of the party was shipwrecked near present-day Galveston. The trip took eight years. "Through that barren country we suffered nearly unendurable hunger," Cabeza de Vaca later wrote. He and his three surviving comrades would have perished just two years into their journey were it not for a group of Indians near the Guadalupe River who resuscitated them with the sweet fruit of the prickly-pear cactus, before providing guidance through the "uninhabited wastes" that lay ahead.

Despite the kindness Cabeza de Vaca's party had been shown, the Spanish official who eventually received them at the outpost of Culiacán, on Mexico's Pacific Coast, listened to his story and then advised him to turn around in order to "restore the deserted, wasted, untilled land by sending to and commanding the Indians in the name of God and King to return to their valleys and tend the soil." Almost four hundred years later, after the region had passed from Spanish to Mexican and then American control, an Anglo journalist named William Smythe called for the same in his 1900 book, *The Conquest of Arid America*. He envisioned "the rapid settlement of western America during the twentieth century. It lies there now a clean, blank page, awaiting the makers of history—the goodly heritage of our people."

In both cases, the arid lands were conceived as a place with no past, and thus limitless potential for the future. For the Spanish farmers and monks who established Albuquerque, Tucson, and El Paso del Norte, the charge was simple: make good on the Spanish Empire's claim to the vast interior of North America where no European had ever set foot. More than a century later, when white Americans founded first Phoenix and then Las Vegas, the plans became baroque. Mere presence was no longer sufficient. No, with sufficient drive, the desert was a place where one could remake their entire life around the pursuit of permanent leisure.

Generations of boosters, land speculators, and hydraulic engineers succeeded in building Phoenix from an archeological site into the fifth-largest metropolis in the country and shaping the remote Las Vegas Valley into a global capital of excess. The two cities represent the apotheosis of the newer, Anglo Southwest, which sought to surpass the established enclaves of Albuquerque, Tucson, and El Paso by embracing a lifestyle completely foreign to the climate, one that could only be achieved via profound industrial force. Doing so meant ignoring or appropriating the Indigenous heritage of the region and forcing its Mexican American and Black communities into the role of subaltern laborers. As Phoenix and Las Vegas grew, the older cities of the Southwest adopted the same white supremist ethos, recasting their multicultural birthright into a tourist fantasia.

For all the exertions of their twentieth-century proponents, the Southwest's cities remain overlooked even as the surrounding landscape maintains a place of privilege in the national consciousness with its magisterial canyons and buttes, its vistas of impossible scale, its shimmering sky. The historical claim that there was nobody in the desert has evolved into the contemporary notion that there are plenty of people in the desert, but no one worth knowing. Sure, the Southwest had accumulated a few cities, but what could possibly differentiate them from the other provincial population centers that dot the American interior? I grew up in Albuquerque, and sharing this fact with folks on the East Coast—where I've been long expatriated—leads almost universally to a blank stare, a bland comment about how hot it must be, or a wistful recounting of their desire to visit Santa Fe. At best, my interlocutor has a grandparent who recently moved to Scottsdale or Summerlin and likes it, they think.

Every once in a while, I get a more illuminating glimpse of the casual disdain so many Americans seem to have for the Southwest. I once found a used copy of Denise Scott Brown and Robert Venturi's groundbreaking study of the architecture of Las Vegas at a bookstore in Cambridge, right around the corner from Harvard Yard. When I brought it up to the young clerk, clad in all black with rimless glasses pressed into his eye sockets, he glanced at the title, *Learning from Las Vegas*, and snarled, "I don't think you can learn *anything* from that place. We'd be better off if it dried up and blew away." Magazines dismiss Albuquerque as "brown, flat, bland, lacking character." Every time I take a reporting trip to Arizona, a family member or friend remarks, "I just don't know why anyone would move there"; similarly, even writers from elsewhere in Texas wonder how anyone could live in El Paso "if not compelled to do so." Whenever they dismiss the cities of the desert—as "real" cities, or even viable places to live—outsiders ignore the deep history of struggle that has led to the present day, of multifarious peoples contending with each other and the unforgiving environment to forge a version of America completely unique to the arid lands. Why is it vital that we understand and learn from the mistakes and triumphs of the people of the Southwest? Because the region already looks like what the rest of America is becoming.

Every year, the United States is getting hotter, drier, more urban, more diverse. By continuing to ignore or abhor the Southwest, people elsewhere risk re-creating its most inequitable and unsustainable aspects. But by learning the region's history and attending to the ways in which it has already figured out how to build dynamic urban centers with scarce resources, the nation might stand a better chance of navigating the destabilizations that the twenty-first century will bring. I fear for the future of America. And I wonder: What lessons might we learn by considering the history of my homeland? In order to see where America is going, we have to look at where the Southwest has already been.

In 1990—the year I was born in Santa Fe—a trio of researchers from Central Michigan University and New York City's Baruch College published a paper articulating what they identified as "The Sunbelt Syndrome." With population growth in America's warmest climes already far outpacing that of the Midwest and Northeast, Edwin Hertz, Terry Haywoode, and Larry T. Reynolds wondered what it was, exactly, that drew so many Americans south and west.

To get an answer, the three researchers interviewed 167 residents of boomtowns in California, Arizona, Texas, and Florida. What emerged from these conversations was a vision of the Sunbelt as a place where "fantasy and reality" held equal weight. The academics argued that the entire region was oriented around "its ever-increasing population, its economic prosperity, its anything-is-possible atmosphere, as well as the free and easy lifestyles associated with glamour, casual sexuality, galloping consumerism, and hedonistic individualism."

Hertz, Haywoode, and Reynolds found these developments troubling, to say the least. "Freedom without values becomes nihilism . . . free floating hedonism and self-gratification become a central core of being as an individual ethic replaces the civic ethic of community," they wrote, after remarking on the region's high divorce rate, sexual permissiveness, and general attitude that children ought to "play a secondary role in lives organized around leisure activities." Ultimately, they concluded, "Without meaning, without purpose, without deep emotion, without social ties, Homo Solis dances into a future of comfort, consumption, and self-centeredness."

The researchers might as well have been writing about Lee Chagra. The favored son of a family of Lebanese immigrants who settled in El Paso, Chagra emerged as the most prominent criminal defense attorney in West Texas in the late 1960s thanks to his special knack for identifying which corners the police had cut to secure wiretaps on his clients. When he wasn't in the courtroom, Chagra would head to Las Vegas to unleash his alter ego, the Black Striker, complete with black leisure suit secured by a golden belt buckle, black cowboy boots made from alligator skin, an ebony cane with a satyr's head, and a cowboy hat embroidered with the word "Freedom" in golden thread. It was in this outlandish getup that Chagra would prowl the gambling floor of Caesars Palace, alternating between blackjack games and rattling craps dice in his fist. The Black Striker persona was as much about betting big as it was playing the part of high roller. In between games, Lee peeled bills off a bankroll and handed them to random tourists, asking, "Do you love me?"

In 1975, Lee's deadbeat brother Jimmy joined the fun, financed by a massive drug score: the importation of fifty-four thousand pounds of Colombian marijuana to the Patriarca crime family in Massachusetts. While most high-stakes blackjack players were then limited to betting only $1,000 per hand, Lee was allowed to wager $3,000, which he did at all seven spots on the table, laying $21,000 at a time. Over at the craps table, Jimmy would stack $10,000 in chips on each number, betting $120,000 to guarantee a payout on almost every roll—but risk a disastrous loss if the dice came up seven.

No matter. For a time, the Chagras were raking in so much dough that blowing $3 million in one summer felt inconsequential. But while every high-rolling Vegas story ends the same way, the collapse of the Chagra family was particularly swift. In 1977, Lee learned that he was being investigated by the Department of Justice for drug trafficking, a revelation that came in the middle of a trial where he was representing one of Jimmy's partners in the Massachusetts operation. Rather than keep his head down, Lee called Caesars Palace to have a Learjet pick him up in El Paso as soon as he got out of court. Over the next forty-eight hours, he burned through all but $10,000 of his hard-earned $250,000 credit at Caesars, then another $250,000 of credit at the Aladdin. By Monday morning, the haggard Black Striker was asking for a new stack of chips worth $5,000 from a blackjack dealer and hearing an unexpected response: no.

Call it the Southwest Syndrome: delusions of grandeur mixing with the pursuit of pleasure to disastrous results, all of it amplified by the extremity of its desert setting. In many ways, it's my ambivalence about the sort of society that would produce guys like Lee Chagra that drove me to leave New Mexico in the first place. When, as a teenager, I went up into the foothills of the Sandia Mountains to look out on the spread of Albuquerque at night—that grid of flickering amber—the sense of immensity clashed with an inchoate claustrophobia, arising from my assumption that all those streets were lined with the tract homes of people who were dull, delusional, dangerous, or dilettantes. Space was everywhere, and yet I felt that I was gazing through a gauze of cottonwood seed. I didn't see possibility out there, just a landscape big enough to vanish me. Nearly every college I applied to was in New York or Boston: *real* cities, I thought; the kind with subways, sidewalks people actually walked on, culture.

It was easy to leave the desert behind, to feel the secondhand glow of importance that comes in working for a glossy magazine in a landmark skyscraper, the ego-stroking satisfaction of a Greenpoint salon. But after spending more than a decade in

Boston and New York, the parochial self-seriousness that comes with global cachet wore thin. At first it was the total ignorance about the Southwest that irritated me—the perpetual need to correct friends and colleagues when they asked me about growing up in Arizona, the rote explanation that it actually snows quite frequently in New Mexico, that the Grand Canyon is so far from Albuquerque you could fit the entirety of Ohio and Indiana between them. Then it was the moments when I was directly informed of the indifference of northeasterners to my homeland, as when I was working on an essay about the literature of the Southwest and asked a prominent book critic if he had any recommendations of novels I ought to read. He couldn't think of a single title.

All of this made me long for the guileless Burqueños I'd grown up with, who were, at the very least, always open to the possibilities of the wider world in a way that someone carrying a "New York or Nowhere" tote bag didn't seem to be. Gradually, though, my exasperation shifted into concern as the intensifying effects of the climate crisis made clear just how high the stakes of this sort of blinkered thinking really are. It took wildfire smoke from Canada turning the sky in New York red for many media members to fully digest the enormous danger that people across the West have been living with for decades, a rainless summer in New England for farmers and policymakers to recognize their vulnerability to drought.

For so many Americans, it is only in recent years that the climate has begun to be understood as a hostile force. To them, I say: Welcome. We southwesterners have never known anything different. It's that recognition that prompted me to return home and devote myself to understanding the deep history of the place that shaped me, the ways that twentieth-century demographic explosion was made possible by the subversion of that ancient heritage, and the contemporary actors seeking to reconnect with fundamental methods of living within the limits of the land in order to adapt to the growing instability of the climate.

By immersing myself in the great cities of the arid Southwest—Phoenix, Tucson, Las Vegas, El Paso, and Albuquerque—I was reacquainted with the region's rhythms, so shaped by its spare ecology. Soon, it became obvious to me just how much the rest of the United States could learn from southwesterners, from the adaptations we've made to endure extreme heat and manage scarce water to our knack for stitching a complex social fabric. But to understand what the Southwest *is*—and, by extension, what lies in store for the rest of the nation—I first had to delve into what it *was*, how its history shaped it into a place that birthed politicians as divergent as Deb Haaland and Barry Goldwater, or minor celebrities as eccentric as the band leader Sue Kim and the private detective J. J. Armes.

The through line between the Southwest's unruly history and its portentous present clarified for me on a visit to Mission Garden, in Tucson. The botanical garden was unusually verdant that August, thanks to the bands of rain that had been pushed from the Sea of Cortez up through southern Arizona for weeks, as one of the strongest monsoons in years rejuvenated the parched Sonoran Desert. Even Sentinel Mountain, which sits just west of the preserve, was coated in a layer of shrubs so thick that they were beginning to infringe on the bright white *A* at its pinnacle—an icon of the University of Arizona, whose students have maintained the 160-foot-tall letter since 1916.

Mission Garden's plots are organized chronologically, so I began in the one devoted to the "First Farmers" and proceeded from there to gardens planted with typical crops from the periods when the Tohono O'odham, Spanish, and Mexicans were the dominant cultures in the region, culminating in a collection of gardens representing the contemporary Tucson Basin.

All the precolonial plots were separated from the Spanish era's expansive orchards of quince, pomegranate, and apricot trees by an acequia, a type of irrigation canal first brought to the region by settlers in the sixteenth century. A fat meadowhawk dragonfly flitted over the water as I crossed the channel, steering clear of a spiny fern where two black-and-yellow spiders had strung their webs. On the west side of the acequia, the ancestral peoples' garden spread out under a branching paloverde and a shady cottonwood; at ground level, squash and corn were planted, along with leafy sacaton cotton, which a laminated pamphlet stated was first domesticated in Central America seven thousand years ago and made its way to what's now southern Arizona around 1000 B.C.E. Along with weaving its downy fibers, the ancestral peoples of the Sonoran Desert cultivated the plant for food—its seeds were roasted and then ground with mesquite beans to form a meal that could be baked into hearty cakes. The O'odham kept the practice alive through the nineteenth century, when Anglo farmers displaced both them and their sacaton plants with higher-yielding Egyptian cotton. Sacaton cotton wasn't reestablished in Tucson until 1984, when Gary Nabhan, an ethnobiologist at the U of A, found six seeds catalogued by the university and shared them with a colleague. "Between them," the pamphlet noted, "they succeeded in growing and preserving the seed."

The O'odham garden featured more sacaton cotton, along with hackberry and Seussian screwberry trees. The entrance to the Mexican garden was an arching arbor decorated with a white flowering vine that smelled like honeysuckle. At the other end of the plot, past some pomegranate trees and a splendid prickly pear, was another archway, this one hung heavy with thick, white squash. Next, I paused at the plots of medicinal plants arrayed around the adobe-brick walls of the preserve. The mingling scents of all the shrubs were as overwhelming as the breadth of an apothecary's pantry: there was verbena, jojoba, white brush, ragweed, and red betony; catclaw, kidney wood, and sand sage. Peering down at all the carefully labeled bushes in their infinite shades of green, I watched a sly gecko scurry past, making a break from a cloud of yellow flowers into the shade of the Mexican skullcap.

Rather than focus on the cash crops that southern Arizona became known for in the early twentieth century—citrus, cotton, alfalfa—the American flank of Mission Garden focused on the diversity of peoples who have come to call the valley formed by the Santa Cruz River home. There was a Chinese garden featuring bitter melons and a trio of tall jujube trees that had just begun to fruit, their purple and white orbs like miniature eggplants or enormous grapes; the pamphlet for this plot explained that many immigrants from rural Taishan settled in Menlo Park, the neighborhood that hosts Mission Garden, in the late nineteenth century, and connected that first foothold for the Chinese community in Tucson to later grocers who set up shop among Mexican neighborhoods on the west and south sides of town, many of whom maintained expansive personal gardens where they grew long beans, water spinach, and gai lan. Next door, the just-added Africa in the Americas plot included crops like okra, sorghum, and cowpeas, and was organized around a shimmering bottle tree.

The historical vision laid out at Mission Garden was a radical departure from the standard narrative of American westward progress, of pioneers carving out a new civilization in a desolate landscape previously populated only by hostile nomads. Instead, it showed how generation after generation of different peoples had each engaged with the Tucson Basin's inherent limitations, learning to adapt to scarcity and utilize a far-from-mighty river to the extent that they could. At Mission Garden, I glimpsed an alternative to the dysfunction articulated in the Sunbelt Syndrome: a model of mutual respect and ecological sensitivity that felt more sustainable than the vistas of endless subdivisions and smoldering blacktop that constitute far too much of the region's cities. I saw the preserve as a manifestation of the old Southwest, contrasted with the newer version that rose as its settlements became hulking metropolises after World War II.

Contemporary America may have been created from displacement and dispossession, but the displaced and dispossessed are very much still here, still constituents of our shared society. To survive in a future where the climate will grow only more unforgiving, the Southwest must determine how to merge its ancestral lifeways into the present. For millions of people to thrive in the hottest, driest corner of the continent, it will be necessary to reconnect with the basic principles of living within one's means and taking only what the earth gives. If we southwesterners can do it, perhaps the rest of America can, too.

PART I

THEY ARE MANY AND KEEP COMING

Tiguex—Kuapoga—Phoenix—El Paso del Norte—Cuk Ṣon

MARROW

Its rank is a mountain and must live as a mountain, as a black horn does from base to black horn tip. See it as you come, you approach. To remember it, this is like gravel.

—LAYLI LONG SOLDIER

At first, Albuquerque felt infinite. Though I was born in Santa Fe, my family moved to the hills on the other side of the Sandia Mountains from Albuquerque when I was five, and most of my childhood was spent tromping through heavy snow and building forts out of piñon branches and desiccated cholla fronds. When my parents took me along for the forty-five-minute drive into town I was shocked by the immensity of the city. Each time we rounded the last crook of the Tijeras Pass through the mountains I was startled anew by how suddenly the rocky, wooded mountainsides of the Sandias fell away in favor of a sweeping vista of glass and pavement that declined steeply into the valley of the Rio Grande before rising back up the other side, where, even if you squinted, it was impossible to make out the line where city ended and desert began.

Albuquerque's form and limits only began to clarify once I started middle school on the other side of the mountains. My friends were mostly scattered across the Northeast Heights, the grid of country blocks that surged out from Albuquerque's historic center by the river to meet the Sandia foothills in the decades after World War II. Through the early years of the 2000s, the epicenter of my universe was Active Imagination, a store on Montgomery and Juan Tabo where I spent hours browsing miniatures or playing games, my mind drifting toward escape. On afternoons when my parents' errands ran long, I'd explore the rest of the strip mall, and whenever I had friends in tow, we'd pool change for a burrito from La Hacienda Express or a treat from the Italian ice place, which the Philadelphian owner had decorated exclusively with Flyers gear.

Once we had driver's licenses, weekend afternoons killing time at Active were exchanged for evenings in the far orbit of the Duke City music scene. My understanding of Albuquerque shifted away from the predictable subdivisions of the Heights to the all-ages venues downtown and the smoky coffee shop patios of Nob Hill, where high schoolers could eavesdrop on University of New Mexico (UNM) students with sophisticated piercings and worn-out copies of Anzaldúa and *Pedro Páramo*. From reading the *Alibi*, the city's alternative weekly, I learned to refer to the city as "Burque," though it would take me years to realize that rhyming that nickname with the word *key*, as I had assumed was proper, was about as Anglo as it got.

By that time, Albuquerque's gravitational pull had worked its magic on my parents as well—much as they preferred the quietude of the mountains, both of them ended up working in the city, which made it hard to justify the long commute, especially in the winter when the Tijeras Pass iced over and semis started fishtailing across I-40. They split the difference by moving us to a neighborhood aptly called the Foothills, a tangle of sparsely populated roads at the city's utmost northeast, within spitting distance of the Cibola National Forest. From our porch, any sense of the city as limitless was immediately punctured. The metropolis petered out at the edge of the barren West Mesa, the dark humps of the trio of long-dormant volcanos called the Three Sisters standing out there beyond the city limits, in open space. At night, the streetlights of the Heights' broad, ramrod-straight avenues descended toward the black wriggle of the Rio Grande, tangling at its banks and then reemerging again on the Westside, wobbling their way into the tenebrous beyond. Most startling was how the lights abruptly dead-ended just to the north of us, the border of the Sandia Pueblo representing an impervious frontier.

I was a white Albuquerque Academy kid, a student at the private school that real-deal Burqueños thought of as a country club. The fact that I even hung out downtown at night was considered scandalous; were I to instead begin frequenting the South Valley, Barelas, Sanjo, or the so-called War Zone, a faculty member would surely intervene. All of those neighborhoods were predominantly Mexican, Black, and Indian, and in the imagination of comfortable Anglos and Hispanics in the Heights the vague threat they represented lacked any of the alluring urban edginess of downtown Albuquerque, Nob Hill, or the Student Ghetto that abutted UNM.

That we would think nothing of avoiding these places while still believing in the myth of the three cultures of New Mexico is, in retrospect, appalling. My entire education, first in public school in the East Mountains and then at the Academy, involved extensive instruction about how the state's special character had to do with the unique comity between Hispanics, Anglos, and Native Americans. Yes, there had been bloodshed in the distant past—the Pueblo Revolt of 1680, the Mexican-American War, century after century of skirmishes with the Comanche, Apache, and Navajo—and, sure, the bipartisan militarization of the border that began under President Bill Clinton and was supercharged after 9/11 was fueling anti-Mexican rhetoric in dreaded Texas and Arizona. But here in New Mexico, the Land of Enchantment, we were immune from such troubles. This was a place that had figured the whole racism thing out, where all could partake in the state's affairs as peers. Bill Richardson was governor; Marty Chávez was mayor. Equality reigned.

This line of thinking was nonsense. But it was nevertheless a gospel that both Anglos and Hispanics believed in—at least the ones monied enough to distance themselves from the depredations experienced by the largely Latino and Indian underclass living without utilities along gravel roads in the South Valley or curled up in blankets and cardboard in bus shelters on Central Avenue.

I was midway through high school when I began to unlearn the fiction of New Mexican racial harmony, part of a summer theater program with a company called Tricklock that held its rehearsals at the National Hispanic Cultural Center in Barelas. The NHCC was a brand-new facility, a tasteful campus of well-lit adobe galleries and cavernous performance venues, the walkways between them lined with flowering ocotillos and leafy cottonwood trees.

Just up 4th Street, the Barelas Coffee House's tables were always crowded with families savoring hulking platters of huevos rancheros and steaming bowls of pozole; the supposed industrial wasteland of South Broadway was in fact full of markets and taco joints; across the river in the South Valley, the streets nearest the Rio Grande's muddy banks were made shady by the Bosque's overhanging trees, and as these routes turned west, they opened out into wide fields and quiet neighborhoods cut through by long-tenured acequias.

These environs were poorer than the Heights, certainly. Walls were tagged up, carjackings common. I was particularly incensed one afternoon after rehearsal when I got back to my car and realized someone had boosted four leather binders stuffed with CDs that I kept on the passenger's seat. Even so, I fell under the spell of the older quarters of Albuquerque, a version of the city distinct from the generic suburbia I otherwise inhabited, where the windows of big-box stores shimmered in the sun, a mirage of plenitude. However derelict, the old neighborhoods made the boosterish claims about New Mexico that I'd ingested more believable: that it was a place of sweeping history, as foundational to the American experiment as the English colonies of New England and Virginia that dominated our textbooks.

I began to understand Albuquerque as a city that operated across a fundamental divide. But that new appreciation only begged more questions. How had the city become so deracinated? What were the forces that conspired to enforce a sense of sameness on a place of such immense diversity? Those questions only began to resolve when I came across a startling bit of information, more than a decade after leaving New Mexico for college, that seemed to have been intentionally neglected throughout my childhood: my hometown had spent more years as two distinct cities—New Albuquerque and Old Albuquerque—than it had being one.

The first convention between two of the groups that would eventually make up New Mexico's three cultures came in 1539, just three years after the bedraggled expedition of Álvar Núñez Cabeza de Vaca reached Culiacán. One of its four survivors was an enslaved Moor from Morocco known as Estevánico, a resourceful polyglot who had learned to communicate with many of the Indians the group encountered on their overland trek. "He was constantly in conversation," Cabeza de Vaca wrote of Estevánico, "finding out about routes, towns, and other matters we wished to know." Meanwhile, the three Spaniards assumed a haughty sense of detachment, hoping to convince the Indians that "we came from Heaven." When the Spanish resolved to send another exploration party into the desert, Cabeza de Vaca was cast aside and Estevánico was commissioned as a guide.

Estevánico led the group to the Zuni settlement of Hawikuh, a large complex of interconnected sandstone rooms built on a high, red-earthed mesa about 120 miles west of the Rio Grande. The urban form of Hawikuh was a legacy of the ancestral Puebloans, the first people to cultivate maize on the Colorado Plateau, a 150,000-square-mile expanse that connects the western slope of the Rockies to the Sonoran and Chihuahuan Deserts. Around 500 C.E., the ancestral Puebloans shifted from living in dugouts and natural rock alcoves to building permanent shelters. Those homes became more elaborate until, by around 1200, they had evolved into vertical cities ensconced in the overhangs of cliffs, most famously at Chaco Canyon and Mesa Verde. Most structures could house dozens of families and the largest, Pueblo Bonita, was a freestanding semicircular complex at least four stories tall, with more than six hundred rooms and forty kivas, the ceremonial gathering spaces that could only be accessed by climbing down a ladder jutting out of a hole in the domed roof.

I knew nothing about the ancestral Puebloans when my parents first took me to visit Chaco Canyon as a kid, but the place made an indelible impression. I remember the intense heat of the sun on the sandy footpath that led through the belly of the canyon, the yellow of the cliff faces on either side, then disbelief at how the surviving walls of a greathouse called Kin Kelso seemed to emerge from the rock itself. Only up close did the stone brickwork of its walls become obvious, even as the material continuity between natural world and human creation remained. Then there was the great kiva at Chetro Ketl, a perfect circle carved into the earth and walled in flagstone, large enough for more than a hundred people to sit comfortably.

In the early 1100s, the Colorado Plateau suffered under a crushing fifty-year drought, followed by decades of climatic instability. This meant that, for all their grandeur, the cliff dwellings were only occupied for about a century. By 1300, all had been abandoned, with many of their former residents heading south and settling along the Rio Grande and its tributaries, where they became integrated into the existing Indigenous population. What resulted were upward of a hundred villages across the region where the old forms of the cliff dwellings were softened by the necessity of building out of mud brick rather than rock. Life in these villages was organized around a new religion borne out of the cultural mingling, wherein ornate Katsina dolls were used to invoke and communicate with a pantheon of spirits. Despite those cultural commonalities, it was a balkanized world, with eight distinct linguistic groups—from Hopi in the far west to Tewa in the north and Piro in the south—and a wide diversity of cultural and religious practices particular to each settlement that emerged over the next two hundred years.

By the sixteenth century, several hundred people lived in the Zuni town of Hawikuh, and its placement gave it an ample view of the scrublands to the southwest over which the Spanish would have approached. Given the importance of gift giving among many of the tribes he had already encountered, Estevánico attempted to smooth his arrival by sending ahead a scout with colorful feathers and a rattle. The Zuni were unimpressed. They warned Estevánico and his men to turn back; when they didn't, most of the group was imprisoned. The last record of Estevánico has him slipping out of his bonds and fleeing over the mesa, accompanied by a hail of arrows.

Despite the ignominious end to the adventure, the Spanish became besotted with the stories of a friar who had traveled with Estevánico named Marcos de Niza. Since the Spanish colonial project was animated as much by the missionary fervor of the crown as by the earthly pursuit of wealth by its subjects, de Niza's claim that Hawikuh was in fact one of several fabulously wealthy cities that could be found in the northern deserts was eagerly received in Mexico. Thus was born the myth of the Seven Cities of Cíbola, places so rich in gold that the metal was used to build walls that glistened invitingly in the sun. In 1540, just one year after Estevánico's failure, Francisco Vázquez de Coronado rode out of Culiacán to conquer the Cities of Cíbola, bringing with him an army of 350 Spaniards, 1,300 Mexican Indians, and 4 Franciscan missionaries.

De Niza led Coronado straight to Hawikuh. Undeterred by Coronado's army—and, perhaps, enraged by the interruption of their celebration of the summer solstice—the Zuni charged. At least ten were immediately killed in the volley of Spanish gunfire, a great crash of explosive force that resounded across the mesa. It was undoubtedly the first time the Zuni had witnessed the violence of firearms. Terrified, they retreated from Hawikuh to their other settlements across the plateau while Spanish soldiers went room to room, grabbing corn and anything else that looked edible.

Despite his immense displeasure at discovering that the golden city de Niza had promised was actually built from stone and mud, Coronado used Hawikuh as a base of operations for several months. While he was there, the conquistador received emissaries from several other tribes who had heard about the Spanish guns and hoped to head off a confrontation. The man who visited from the village of Cicuye (later known as Pecos) was labeled "Bigotes" in recognition of his impressive mustache. Bigotes was a savvy diplomat, well acquainted with the thriving intertribal trade that characterized life in the arid lands before European contact. He seems to have even spoken Nahua, allowing him to converse with some of the Mexican Indians in Coronado's army, who could in turn translate for the conquistador. To charm the new arrivals, Bigotes bore an exotic gift: robes made from buffalo, the hides surely acquired from one of the nomadic tribes of the Great Plains. When the Spanish asked where they could find the enormous "prairie cattle" Bigotes described, he said he'd be happy to show them, though it would mean traveling far to the east.

Coronado put the scouting party that would ride with Bigotes under the command of Hernando de Alvarado, an artillery captain who had grown up in the northern Spanish province of Santander and first crossed the Atlantic at the age of thirteen. Still only twenty-three, Alvarado had become a favorite of Coronado's for throwing his body in front of the general after he fell off his horse during the storming of Hawikuh.

Alvarado followed Bigotes toward Cicuye. The sagebrush plain seemed infinite, until, imperceptibly, the declination of the Rio Grande Valley began. At last, the verdant ribbon of the river began to reveal itself beneath a magisterial mountain ridge. As the sun faded and Alvarado, his men, and their Indian guides made camp, the color of the ridge kept changing, from pale green to ruddy bronze to the vibrant pink of watermelon flesh. What else could the Spanish call the mountains but Sandia? Eventually, the light failed entirely, leaving behind a monumental silhouette against the starlight.

It was early September when Alvarado met the river, and the banks were well greened by the annual monsoon season that crests in late summer. This, Bigotes explained, was the province known as Tiguex. It was a place where Indigenous villages had stood for at least five centuries. Alvarado later wrote that the river "flows through a broad valley planted with fields of maize and dotted with cottonwood groves. There are twelve pueblos, whose houses are built of mud and are two stories high." Of the people occupying those adobe homes, Alvarado wrote in seeming amazement that they seemed "more devoted to agriculture than war."

Indeed, the Tanoan-speaking Tiwa were concentrated in as many as sixteen villages that stretched for thirty miles from Isleta in the south to what's now Bernalillo—the same territory currently occupied by Albuquerque. The village in the foothills of the Sandia Mountains, and which took the same Spanish name, was among the largest of these settlements, with a population of around three thousand. All told, the 110 or so settlements across all of modern New Mexico and Arizona are thought to have been home to some eighty thousand people when the Spanish first arrived. The Tiwa, like the Tewa, Piro, Keres, and Hopi who lived in the other villages, were hardly startled to encounter a foreign culture: the nomadic Apache and Navajo had arrived in the region around 1400 C.E., though their regular raids quickly made them blood enemies of the sedentary, farming peoples. The Spanish word for "village," *pueblo*, became the name with which the settlements would forever be associated in the European imagination.

Despite the warm reception that Alvarado found in Tiguex, the Tiwa were hardly prepared to continue provisioning the Spanish after Coronado decided it would be better to spend the winter among them than out in the ruins of Hawikuh, where the bulk of his army had remained camped since the summer. Coronado's second-in-command was García López de Cárdenas, a veteran soldier who had served in Italy and the Caribbean before reaching Mexico, and he was sent ahead to make the necessary arrangements. Initially, Cárdenas set up a camp for the army outside of the existing pueblos. But as the days grew shorter and the nights colder, Cárdenas requested that one of the pueblos be vacated. Begrudgingly, the Tiwa living in a village the Spanish called Alcanfor assented to the displacement, scattering across the valley to stay with kin elsewhere.

Tensions only grew once the mountains became dusted with snow. As the days grew colder, Spanish soldiers burst into other pueblos to ask for the blankets that the Tiwa wore around their shoulders; whenever they were met with confusion or refusal, the soldiers tore the blankets away by force. Meanwhile, when the thin vegetation on the plain that extended toward the river from the Sandias withered away to nothing, the Spanish set their horses and cattle loose on the Tiwa's cornfields, where they chomped down what remained of the harvested corn stalks—an important source of fuel for the pueblos in the winter.

The hospitality of the Tiwa reached its limit in late December, when a man named Villegas led a group of soldiers into one of the pueblos. The raiding party confiscated chickens and warm clothing while Villegas sexually assaulted a woman. The next day, a group of Tiwa sought vengeance by descending on a field where many of Coronado's horses and mules were pastured, killing a guard and spooking the animals, which fled across the frozen valley. Furious, Cárdenas marched to the pueblo of Arenal with sixty men. The entrances to the village had all been blocked, but he could hear a massive racket inside: the Tiwa had trapped some of the Spanish horses and were delightedly chasing them around a central plaza while onlookers hooted at the spectacle. Cárdenas demanded that the horses be returned but was drowned out by the carousing.

Cárdenas resolved to smoke out the insolent Tiwa. He ordered his men to batter holes into the adobe walls, then light smudge pots and throw them inside. As black smoke billowed through the mud brick corridors of the pueblo, the people inside ran outside to find fresh air, where they were met with Spanish guns. All who escaped the first volley of the footmen were pursued by mounted soldiers. Once the pueblo had quieted, the Spanish filed in and yanked all of the stragglers outside, where they were tied to stakes.

In desperation, a handful of Tiwa laid down their weapons and formed a cross with their hands. Looking on with a stony countenance, Cárdenas met each gesture with his own sign of the cross to indicate that he understood the intention to surrender. A nearby soldier cautioned Cárdenas that the church might frown on his behavior, exhorting him, "Do not show them the cross unless you expect to fulfill your promise." Cárdenas said it didn't matter: he hadn't performed a proper genuflection, and thus the sign was meaningless. He ordered that the Tiwa who had surrendered be rounded up into a tent. When they saw the Spanish making preparations to set fire to the canvas, the Tiwa panicked, grabbing stones from the ground to use as weapons. "They lanced and put the sword to everyone who was there defending themselves," one soldier recalled. Seeing that no Tiwa would be spared, the remaining Spaniards "set fire to the others, who were still tied to the posts, and burned them alive."

At least 130 were killed in the Arenal massacre, but that only marked the beginning of the Spanish assault. The Spanish expected total obedience from the peoples whose lands they hoped to scour for wealth. All resistance was met with retribution justified by the idolatry and, to the Spanish, "senseless" violence of their victims. Ten of the pueblos in Tiguex were burned to the ground in the ensuing months. Spanish soldiers paused only to recover any wood they could find to warm themselves back at Alcanfor. Almost all of the Tiwa fled north to the pueblo of Moho, near where the Jemez River meets the Rio Grande. There they endured another brutal siege. By the time the Spanish seized Moho, springtime had arrived and Coronado shifted his attention back to the pursuit of gold, which Bigotes—perhaps eager to steer the murderous strangers out of the pueblo lands—assured him could be found in the east, out with the buffalo. Blackened and brutalized, the province of Tiguex was abandoned. For the Tiwa who did resettle there, the vicious men in metal armor became a matter of story and memory—it would be sixty years until they again faced the Spanish.

The ten pueblos sacked by Coronado's army remain largely uncommemorated in Albuquerque. Most of them lay buried beneath the contemporary streetscape. There is an Arenal Road that cuts through the South Valley, running past a handful of fallow yellow fields and a succession of little beige houses fronted by cinder block walls or chain link fence. But the real Arenal was probably twenty miles north, near Sandia Pueblo, one of only two original settlements in the area that survived the Spanish conquest. On the opposite bank of the river from Sandia, the remains of one pueblo have been preserved: Kuaua, which means "evergreen" in Tiwa. There, the National Park Service has partially restored the floor plan of the pueblo to showcase its geometric pattern of rectangular rooms, oriented around a wide square plaza with a kiva in the center. It is called the Coronado Historic Site.

Despite what government officials and tourist bureaus claim about New Mexico's three cultures, it's been the Spanish who are memorialized most often and most loudly—none more so than the criollo aristocrat Juan de Oñate. Oñate was the son of the Basque conquistador who founded the city of Guadalajara and then became fabulously wealthy by mining silver near Zacatecas. As befitting a child of privilege, Oñate married well—to the granddaughter of Hernán Cortés himself. In the last years of the sixteenth century, it was Oñate who was chosen to make good on the promise of Coronado's exploratory mission with the establishment of a permanent colony in the region. The assignment had urgency to it: Englishman Sir Francis Drake had recently mapped the coast of California, raising fears among Spanish authorities that their European rivals might challenge a colonial claim that still existed only on paper.

Just after the new year in 1598, Oñate gathered his party in the town of Santa Bárbara, nestled in the mountains of southern Chihuahua. Under his command were five hundred soldiers and settlers—not significantly larger than Coronado's army, but they would be moving much more slowly, because the settler families dragged eighty-three wagons and seven thousand cattle, sheep, donkeys, and horses with them. Prepared to threaten war on any tribe that opposed him, Oñate also brought eighteen barrels of gunpowder for his soldiers. The settlers rode down into the comfortless plain, with the Sierra Madres at their back and little understanding of how far they were from the Rio Grande. The massive party tromped north, the hazy blue shapes of the distant mountains ahead never seeming to rise above the horizon line. The constant, brutalizing sound of the wagon train—bleating goats, shuddering axles, yelling drovers—was a cacophonous soundtrack to a view that only ever got more ominous. At first, it was mile after mile of yellowed gramma grass, interrupted by the sinewy trunks of an occasional soaptree yucca quivering beneath the sun. But after months of marching, even that limited vegetation boiled away as the vast sand dunes of northern Chihuahua swallowed the world.

On April 19, with the northern mountains beginning to coalesce into tangible shapes, Oñate dispatched a small advance party. He was running short of water and was desperate to find the river previous expeditions insisted lay ahead. After four days of hard riding and rationing water, the group finally began to feel the earth subsiding into a valley. Then, they saw green. By the time the scouts actually reached the Rio Grande, their horses were so thirsty they bolted into the muddy river, which was running high with spring snowmelt. Two of the beasts were swept away and never seen again, but the settlers hardly noticed. One later recounted how "our men, consumed by the burning thirst, their tongues swollen and their throats parched, threw themselves into the water and drank as though the entire river did not carry enough to quench their terrible thirst." Once they were satiated, the Spaniards collapsed under the shade provided by a few cottonwood trees. When Oñate reached the advance scouts with the rest of the wagon train a few days later, a great celebratory feast was held, with fish and meat roasted on a bonfire.

Once the group had recovered from its 350-mile passage, Oñate ordered his soldiers to find the best spot for fording what the Spanish were then calling El Rio del Norte, then build a chapel there. On April 30, the Franciscan friars traveling with the settlers held a mass, after which Oñate claimed for the Spanish crown all of the land drained by the river—never mind that no Spanish explorer had yet made it anywhere close to its headwaters in the mountainous homeland of the Ute people in today's southern Colorado.

As Oñate headed further north, it became clear that the sixty years that had passed since Coronado's journey had done little to reduce Puebloan fear of the Spanish. Practically every village Oñate encountered was empty, its denizens choosing to flee rather than parlay. Without delay, Oñate proceeded all the way to the confluence of the Rio Grande and the Rio Chama. This was the spot he had already decided to set up as the capital of the colony of Nuevo Mexico. When the party reached the fork in the river, they found a deserted pueblo known as Yunge Oweenge. They built over the ruins, constructing a Catholic church and some provisional shelters out of the rubble, which they dubbed San Gabriel.

With a headquarters established, Oñate began to explore his newly claimed dominion. According to Spanish histories, Oñate's initial encounters with the various pueblos he visited—Taos in the far north, the thickly populated Galisteo Basin, villages of the Jemez Mountains—were cordial. Only when he reached Acoma did the pattern that Coronado had established at Tiguex repeat itself: the Spanish met the accommodating Puebloans with imperious demands of ever-escalating force until they dropped the charade of diplomacy entirely.

Acoma rises from the top of a high butte in the badlands north of Zuni, a magnificent Sky City first built in the eleventh century, giving it a claim to being the oldest continually inhabited town in the contemporary United States. Oñate set up a camp at the base of the butte in October 1598, the adobe buildings at its summit barely visible above the vertiginous sandstone. A group of Acoma descended to meet the Spanish, then graciously showed them how to navigate the 350-foot climb to reach their village. At the top, the Spanish were amazed to see buildings of three and four stories, organized around deep, natural cisterns that stored drinking water accumulated from snowmelt and the thunderstorms that whipsaw over the region throughout the late summer and early fall. With the Acoma warily looking on, a few soldiers stood at the edge of the butte and admired the panoramic vista, firing their guns as a salute to the Spanish who remained in camp below.

After the brief visit, Oñate continued to the west, hoping to find his way to the sea. Two months later, another party of conquistadors arrived at Acoma, led by Oñate's nephew, Juan de Zaldívar. Leading a few men up to the top of the butte, Zaldívar asked for provisions. This was early December, and the Acoma were incredulous that the Spanish would make such a request just as winter was coming on, but nevertheless offered the visitors cornmeal. Zaldívar demanded more. What transpired next is a matter of much historical debate: by one account, an impatient soldier named Vivero snatched two turkeys; by another, an Acoma woman was sexually assaulted. In any event, a melee ensued, during which the Acoma killed Zaldívar and ten of his men.

When Oñate heard from the survivors what had happened, he ordered a retaliatory strike. The Acoma's natural defenses and the bravery of their soldiers proved no match for the Spanish cannons and guns: in the end, eight hundred were killed and another

six hundred taken into custody. Oñate ordered that all male prisoners over the age of twenty-five were to have one foot amputated, and that both they and the pueblo's women be forced into
servitude. Two Hopi who happened to be visiting the Sky City
during the assault were also apprehended. A hand was amputated
from each man before they were told to return to their villages
as a warning.

Oñate's retaliation against the Acoma was so disproportion-
ate that when the viceroy of Nueva España heard what he'd done,
the conquistador was immediately recalled to Mexico City. He
would eventually be tried and found guilty of violating the strict
Spanish regulations over how to treat Indigenous subjects—just
as Coronado had been. Nevertheless, Oñate had fulfilled his
mission: for the first time, a permanent Spanish population had
taken root in Nuevo Mexico.

In 1610, the governor who replaced Oñate, Pedro de Peralta,
moved Nuevo Mexico's capital from San Gabriel twenty-five
miles south, to a more defensible site along a little creek in the
foothills of the Sangre de Christo Mountains. There, ruins of
another abandoned pueblo had been found, a village once known
as Kuapoga, or "the place of the shell beads near the water."
Now, it became La Villa Real de Santa Fé. The settlement was
initially conceived as a military presidio fortified to defend the
Palace of the Governors, a one-story administrative building
made from adobe, with a simple wooden colonnade, or portal,
facing a central plaza. As the Palace of the Governors was being
built, the first Virginians were abandoning Jamestown and the
pilgrims were still a decade away from setting sail for Plymouth.
It remains the oldest governmental building in the United States.

Within twenty years, a thousand settlers had journeyed to
Santa Fe, almost all of them Mexican mestizos rather than pure-
blooded Spanish criollos. As the small outpost continued to grow,
Franciscan friars fanned out across the pueblos, constructing
churches and baptizing all the Indians they could. In some cases,
the presence of the Spanish was welcomed, particularly among the
villages at the frontier of the Puebloan homelands that had long
suffered from Apache and Navajo raids. But whatever protection
the Spanish extended to the pueblos had to be weighed against

the imposition of the encomienda and repartimiento systems that animated Spanish colonial rule. Under those regulations, Spanish settlers were entitled to seize supplies and force Indians to work land they had been granted by the crown. Despite Madrid's official prohibition against slavery, many Puebloans were kidnapped from their families as children and grew up laboring in Spanish households, often never learning the language or traditions of their people, or even where they had been born. These enslaved Puebloans became known as Genízaros, a name that cordoned them into a murky ethnic space—not mestizos, yet not culturally Indian, either.

Many Franciscans took their own Genízaros, whom they impressed into servitude at the churches that complemented the encomienda system's attempted assimilation of the pueblos into Spain's imperial economy by stamping out traditional religious practices. These missionaries labeled Katsina rituals as a form of devil worship that was incompatible with Christian faith, and punished Puebloans for practicing their ceremonies: at one of the Hopi missions, a priest named Salvador de Guerra whipped an accused apostate until his back was drenched in blood.

Whatever kindness most Puebloans might have been willing to extend to the Spanish evaporated in the early 1640s, when a smallpox outbreak and drought hit at the same time, killing thousands. Resistance to the Franciscans—whose God had so clearly failed to provide for the Puebloans—grew from there, and each time a Puebloan questioned a missionary or asserted their own faith, they were punished. Governor Fernando de Argüello Carvajál had forty Puebloans flogged and then imprisoned for their supposedly seditious beliefs, then ordered that twenty-nine leaders of the Jemez Pueblo be hanged for supposedly colluding with the Apache and Navajo. Things only degenerated further when an even more severe drought struck in the 1660s. So many Puebloans died—whether directly killed by a Spanish colonist or victim to disease or starvation—that their population fell to just twenty thousand, a quarter as many as had occupied Nuevo Mexico when Coronado first arrived more than a century before.

Despite the excesses of the Spanish, it was only in 1675 that the leaders of the Pueblos resolved that collective action was necessary. That year a newly appointed governor, Juan Francisco Treviño, attempted to cement his authority by arresting forty-seven spiritual leaders on charges of practicing witchcraft, mostly from the Tewa-speaking northern pueblos. Three were hanged, the rest tortured and imprisoned in Santa Fe. Incensed, a large group of Tewa traveled to the city, refusing to leave until Treviño released the captives. Apparently surprised by the scale of the protest, Treviño relented and let everyone go.

One of the members of the group who was allowed to return home was a man named Po'pay, from Ohkay Ohwingey, the nearest pueblo to Santa Fe. In his forties at the time of his arrest, Po'pay had lived his entire life under Spanish rule, covertly participating in traditional dances and ceremonies to avoid the wrath of the friars. The Ohkay Ohwingey anthropologist Alfonso Ortiz suggested that Po'Pay's name, which translates as "ripe cultigens," indicated his rarified place in the pueblo. "Because the name is generic, and therefore, potent," Ortiz wrote, "it could not be conveyed on an infant. Given the Tewa scheme of things about naming, so potent a name had to be earned, and the only appropriate channel through which one could earn such a name is through religious service after adulthood." Likewise, the harvest associations of Po'Pay's name imply that he was an influential part of the summer moiety of his pueblo's religious practitioners (with the other spiritual leaders participating in the symmetrical winter moiety), "perhaps even the chief priest."

In the months after his release, Po'Pay and the other religious leaders began meeting in secret to discuss how they could loosen the grip of Spain over their homeland. The group of conspirators probably included Diego Xenome from Nampe, Nicolas de la Cruz Jonva of Po-sogeh, and Domingo Naranjo, from Ka-'p-geh. Most of the group spoke Tewa, either a northern dialect or one from the Galisteo Basin, though Keres speakers from Cochiti and San Marcos also attended. Gradually, war captains from further afield began to appear, including Luis Conixu from Walatowa (in the region the Spanish called Jemez). Each leader was given ample time to speak, such that no individual dominated the proceedings. Still, Po'Pay emerged as the participant best able to articulate the need for action. According to Puebloan oral tradition, the other leaders believed that he had "the cunning of a fox and the determination and heart of a bear."

The secret meetings culminated in a large nighttime gathering in the summer of 1680, with the visitors seated in deference to the host council, who were positioned closest to the great kiva's smoldering fireplace. After the hosts offered a prayer, one visitor suggested the need for a singular leader who might oversee a war against the Spanish, with a consensus quickly forming that Po'pay was the obvious choice. Po'pay stood, thanking the others for putting their trust in him, then asked each participant to determine which villages could be relied upon to join the rebellion and which were thought to be loyal to the Spanish. Next, the gathering discussed timing. Given that Santa Fe would soon be receiving a wagon train of supplies from Mexico, it was necessary to strike quickly. Likewise, their best chance of success required all of the pueblos to act at once, preventing the Spanish from organizing a defense with their superior arms. To that end, it was resolved that runners should convey cords made from yucca root to every sympathetic pueblo, each one with a particular number of knots worked into it. Every day, one knot would be removed from the cord. When none remained, the pueblos struck.

On August 10, the colonial order that the Spanish had established over the preceding eighty years fragmented. At Sandia, one rebel used an ax to hack away the arms of a figure of Saint Francis, then gathered dry straw inside the choir loft and lit it on fire. At Jemez, a priest named Juan de Jesús was forced to crawl naked through the mud while a Puebloan rode him like a donkey; 270 miles west, a crowd of Hopi hung a missionary named Joseph de Trujillo from one of the pine vigas protruding from the settlement's church, then threw his body on a pyre that included church documents and wooden effigies of Catholic saints. All across the vast sweep of the Pueblo lands, nearly every village was caught up in the choreographed assault, which the Spanish experienced as a swell of inexplicable violence, as terrifying as it was humiliating. Meanwhile, Po'pay and the other faith leaders remained inside their respective kivas, praying on the Katsinas the Spanish had sought to displace with their singular God.

After the churches had been burned and twenty-one friars

were dead, along with more than three hundred lay settlers, the rebels turned their attention to Santa Fe. On August 15, an army of five hundred Pueblo soldiers approached the presidio. The governor, Antonio de Otermín, invited a Puebloan whom some of the soldiers recognized inside the walls to negotiate, a Tano man who had learned to speak Spanish and been given the name Juan. He advanced to the Palace of the Governors, wearing looted armor and carrying a Spanish gun, and stood, unfazed, as Otermín demanded to know why the Pueblos had revolted. Instead of an answer, Juan offered an ultimatum: either the Spanish could return to Mexico, or they would be slaughtered.

Otermín refused to uproot the nine hundred or so residents of Santa Fe, instead proclaiming that they would weather the siege. The Puebloans choked off the Santa Fe River, the creek that fed water into the city, as more and more reinforcements arrived, until some 2,500 people—warriors as well as their families—were massed around the city. Conditions quickly grew intolerable inside Santa Fe; animals died of thirst and civilians screamed for salvation. On August 19, Otermín organized a desperate surprise attack. A force of one hundred soldiers threw open the gates in the early morning: though they were badly outnumbered, the firepower of the Spanish enabled them to fend off the unprepared mass of Pueblo warriors. Otermín was severely wounded by a gunshot to the chest. While the Puebloans regrouped, the Spanish loaded wagons with everything they could and raced south, hardly stopping until they reached the safety of the Franciscan missions in El Paso del Norte.

The Pueblo Revolt stands as one of the few moments in American history when a network of Indigenous communities managed not just to hold European colonizers at bay, but to fully jettison them from their homeland. Despite the watershed nature of the rebellion, it only succeeded in suspending Spanish domination for a dozen years. And those years were not happy times: Apache and Navajo raids continued, as did bouts of disease and drought that enervated the pueblos. The absence of a common enemy also allowed for old divisions to reemerge, leading to internecine skirmishes that further sapped Puebloan resources. By the time Diego de Vargas led a new army up the Rio Grande from El Paso del Norte in 1692, the Puebloans had been so decimated that he managed to reenter Santa Fe without firing a shot.

For decades, de Vargas's reconquest of Nuevo Mexico was commemorated every fall during the Fiesta de Santa Fe in a reenactment dubbed the Entrada. In the standard version, a prominent Santa Fean played the role of de Vargas, striding onto Santa Fe Plaza wearing a golden conquistador's helmet and a purple cape. As a crowd looked on, he removed his helmet and sword to signal his peaceful intentions and greeted an official from one of the nearby pueblos. Both men then directed their gaze to a doll of the Virgin Mary propped on a wooden pole. Reassured, the pueblo leader invited de Vargas to the stage, where all the reenactors prayed together, their shared faith bridging the cultural divide.

In truth, de Vargas's Reconquista was just as deadly as the forays of Coronado and Oñate. Despite the absence of violence when de Vargas first reentered Santa Fe in 1692, the Puebloans who had taken up residence inside the walls of the city refused to leave when he returned with settlers a year later. A bloody battle followed. After de Vargas prevailed, seventy captive Puebloans were executed by firing squad. The forum for their punishment was the same plaza where, hundreds of years later, the Fiesta de Santa Fe, an elaborate fiction of peace and equanimity, would be staged.

Reflecting on the pageant, Santa Fe mayor Javier Gonzales, who had once played de Vargas, said, "As proud as I was to participate in this important community tradition, I do believe it's time that we be truthful about the actual events that occurred during resettlement." Though de Vargas was indeed a man of faith, "force was still used to resettle Santa Fe and the indigenous people were forced to adopt Christianity as their religion. To imply something other sends the wrong message that the Spaniards were welcomed." In the years after Gonzales's call for a reevaluation of the Entrada, it was disrupted by protests, leading to the pageant's being canceled entirely in 2018. Afterward, an activist from the San Ildefonso Pueblo named Jennifer Marley, who had been arrested in one of the earlier protests, told the *Santa Fe New Mexican* she was happy that the tradition was "finally getting laid to rest." At the same time, Regis Pecos, the former governor of the Cochiti Pueblo, sought to reassure Santa Feans that organizers hoped "the original intent of celebrating the shared faith" would remain central to the annual fiesta.

Though many contemporary Puebloans are practicing Catholics, the idea that the faith was widespread after the Pueblo Revolt is yet another matter of historical revisionism. After seizing Santa Fe, de Vargas embarked on a grim, two-year campaign to bring the rest of the pueblos to heel. Many people fled their homes rather than submit to a forced baptism. Not that fleeing guaranteed them safety or the ability to continue their Katsina traditions; in the summer of 1694, the Spanish stormed a cliff in the Jemez Mountains where several hundred Puebloans were in hiding, their only defense stones they had collected from a riverbed that they rained down on the approaching soldiers. There were no Spanish casualties in the attack, but eighty-four Jemez were slain and more than three hundred others taken prisoner. A Franciscan priest who was traveling with the army baptized two of the combatants shortly after they were apprehended. Newly Christian, the men were promptly executed on the cliff, the report of their killers' gunshots dampened by the juniper trees.

Once Spanish control of Nuevo Mexico was secure, only around fourteen thousand Puebloans were left. So many had perished in the intervening century that many of the historic villages were abandoned, the population consolidating in order to survive. The six Zuni villages that had existed before the arrival of Coronado became one. Of the sixteen towns that once composed the old Tiguex province, only the pueblos of Isleta and Sandia remained. Yet despite their diminished numbers, fears of provoking another revolt by the Puebloans fundamentally shifted Spanish colonial policy following de Vargas's Reconquista. The Puebloans were treated more as military allies than as subjects: so long as they expressed some deference to the Catholic Church and could be counted on to take up arms against the Navajo, Ute, and Apache, they were largely left to their own devices. Only two pueblos may have remained in Tiguex, but they *did* remain, gradually integrating into the social fabric of the middle Rio Grande Valley even as more and more settlers arrived from Mexico.

Bernalillo was one of the first Spanish towns to take root in the area after the Reconquista, just across the Rio Grande from the Sandia Pueblo. It was founded in 1695, with Atrisco following in 1703. Alburquerque itself was established in 1706 on the eastern shore of the Rio Grande. The new villa was named in honor of the ducal title of Francisco Férnandez de la Cueva, then the viceroy of Nueva España—an attempt by the newly appointed governor of the colony to curry favor with his supervisor. Per the Spanish crown's prescription for all New World settlement, Alburquerque's first building was a church, the forerunner of San Felipe de Neri, on a plaza freshly cleared from the forest of cottonwood trees that hugged the river's bank. "I do not doubt, very excellent lord, that in a short time this will be the most prosperous Villa," Governor Francisco Cuervo y Valdés wrote to the viceroy in Mexico City, "because of its great fertility and for my having given it the spiritual and temporal patrons that I have chosen: namely, the ever glorious Apostle of the Indes, San Francisco Xavier, and Your Excellency, with whose name the town has been entitled Villa de Alburquerque de San Francisco Xavier del Bosque." Over the next seventy years, the same pattern repeated itself across the northern reaches of Nueva España, as missions were established at San Antonio, Tucson, San Diego, Los Angeles, and San Francisco.

Cuervo's sycophantic grandiloquence evoked the romantic notions of "old New Mexico" that have long enraptured Anglos and Hispanics alike, many of whom indulge in a shared fantasy of the colonial era as a prelapsarian world where comely doñas gamboled about the estates their princely families established along the Rio Grande. In reality, Nuevo Mexico was among the most destitute provinces in all the Spanish Empire. Some of its decay was reflective of colonial geopolitics—Madrid was occupied by wars in Europe that saw its influence and wealth diminish in the face of the rising British and French—but the local reality of being a thinly populated domain of shepherds and subsistence farmers surrounded on all sides by hostile tribes also took its toll.

Even as the pueblos had been pacified by the Spanish, in the eighteenth century Nuevo Mexico effectively become a client state of the Comanche, the nomadic tribe who rapidly expanded from their homeland in the Sangre de Cristo Mountains to become the dominant force in the southern Great Plains. New Mexico was caught in the crossfire between the warring Comanche and Apache in the 1710s and '20s, but even after the Comanche and their Ute allies had firmly established their hold over the hinterland of the colony, villages along the Rio Grande became their favored prey. Those raids peaked in the 1770s, when the Comanche sphere of influence stretched from the Rocky Mountains all the way to San Antonio. When the missionary Juan Agustin Morfí visited Albuquerque in 1778, he reported that, though "the land is fertile," it largely lay fallow because the few hundred settlers who lived there were so fearful of the Comanche that they "dare not go out and work the land." The raids were organized to seize food, horses, and people, with the Comanche abducting children and women from poorly defended pueblos so they could be sold at the markets in Taos, ensuring a steady supply of Genízaros for Spanish merchants and ranchers.

The peace that the Spanish eventually brokered with the Comanche stabilized things for a time, but what followed was three decades of geopolitical chaos. Mexico won its independence from Spain, the Texas Republic broke away from the new nation, and then the United States invaded—through it all, Mexico City had little capacity to attend to the needs of a poor province fourteen hundred miles away. The result was that, through the first half of the nineteenth century, the flags flying over Santa Fe kept changing even as the desperate circumstances of the settlers and pueblos remained as they had been since the Reconquista, with only a handful of Hispano gentry managing to become wealthy through trade with the ascendent United States. As the friar Francisco Atanasio Domínguez put it, the inhabitants of Nuevo Mexico were "miserable wretches . . . tossed about like a ball in the hands of fortune." To his point, General Stephen Kearney seized Albuquerque without a fight at the outset of the Mexican-American War in 1846. Even after the Treaty of Guadalupe Hidalgo turned over the northern half of Mexico to the United States two years later (an area that includes today's California, Nevada, Arizona, New Mexico, Utah, and Colorado), Alburquerque remained an afterthought. In 1853, one visitor remarked that the city had "a rather ruinous aspect"; the same year, an American general called it "the dirtiest hole in all New Mexico."

For the new American authorities, the most pressing issue in its newly seized territory was the seemingly intractable conflict between the settled population and the Navajo. With the Comanche preoccupied with fighting their new Anglo antagonists in Texas, Nuevomexicanos had begun organizing their own slaving parties, which rode into Dinétah, the Navajo homeland, to kidnap whoever they could find. The Diné would respond in kind, assaulting villages and killing dozens in search of their lost children and wives.

The Americans chose to side with the settled Nuevomexicanos. After all, the Genízaro trade was a close cousin of southern chattel slavery, and perhaps seemed comparatively compassionate given that Genízaros were sometimes able to serve out the terms of their forced service and live freely, founding towns like Abiquiú, Las Trampas, and Belen. Still, the number of Genízaros was staggering. In the late eighteenth century, one in three Nuevomexicanos are now thought to have been enslaved people or their descendants.

Authorities hoping to transform New Mexico into a stable destination for Anglo settlement used the threat of enslavement to rein in the Navajo, who controlled practically all of the enormous Colorado Plateau. As one Army captain wrote, "The most effective rod of terrorism to be held over these people is the fear of permission being given to the Mexicans to make captives of the Navajos and retain them." In fact, no permission was needed: the U.S. Army tacitly encouraged the continuance of slave raids for the next decade and a half, riding against the Diné whenever they retaliated.

Only after General James Carleton successfully repelled the Confederate army's invasion of New Mexico in 1862 did he resolve to force the territory's nomadic peoples into submission once and for all. The next year, after imprisoning four hundred

Apache at Fort Sumner, Carlton ordered the dismantling of Dinétah: American soldiers burned hogans, slaughtered cattle, and incinerated cornfields. Even the five thousand peach trees that lined the Little Colorado River at the bottom of Canyon de Chelly, the spiritual origin point of Diné culture, were felled. All Navajo were expected to surrender at a military outpost near the canyon called Fort Defiance, after which they would be escorted three hundred miles west, to an uninhabited stretch of the Pecos River known as the Bosque Redondo because of its circular grove of cottonwood trees.

"Say to them—go to Bosque Redondo, or we will pursue and destroy you. We will not make peace with you on any other terms," Carlton instructed his men. "This war shall be pursued until you cease to exist or move." One in five of the Navajo and Mescalero Apache who surrendered at Fort Defiance died on the three-hundred-mile trek east and during their subsequent internment. The alkaline groundwater at Bosque Redondo was nearly undrinkable, and when the prisoners attempted to grow corn their crop was decimated by cutworm, forcing them to rely on meager Army rations, since most of the Army's largesse was funneled east to the ongoing war against the Confederacy.

Only in 1868 were the Navajo able to return to their home-land on the Colorado Plateau. The column of wagons rolling west is said to have stretched for ten miles. After crossing the Rio Grande, the Diné caught their first glimpse of Tsoodził, what the Americans called Mount Taylor. A dormant volcano more than eleven thousand feet tall, Tsoodził is the most southeasterly of the four sacred mountains that delineate the traditional lands of the Navajo. At first, the survivors were seized by disbelief that home could be so close at hand. "When we saw the top of the mountain from Albuquerque," one of the tribe's leaders, Man-uelito, remembered, "we felt like talking to the ground we loved it so."

The deprivations faced by the Navajo over the latter half of the nineteenth century were indicative of the upheaval that spread across the region after the United States declared war on Mexico in 1846. Though the pretense for the invasion was Mexico's refusal to recognize Texas as anything but its own territory, the war served as a way for James Polk to make good on the expansionism espoused during his presidential campaign, when he'd promised to transform the United States into a continent-spanning empire. While the enormous territory seized from Mexico—which spanned from the Gulf Coast to the Pacific—looked good on a map, the land itself was otherwise an after-thought. In the speech Polk gave to Congress following the war's conclusion in 1848, he went long on California's strategic value and the gold that had been discovered earlier that year at Sutter's Mill, but spared only a few lines for New Mexico, referring to it mostly as "the intermediate and connecting territory between our settlements and our possessions in Texas and those on the Pacific Coast."

Many of the earliest Anglo arrivals to the new territory were likewise focused on reaching California, not the land they would have to cross to get there. That included Franz Huning, a German who emigrated in 1848, the year the Treaty of Guada-lupe Hidalgo was signed. Huning had grown up on a farm near Hanover with twelve siblings, more interested in books than the cattle under his charge. He would later recall long hours of read-ing while "stretched out under an oak or a beech, oblivious of everything else but the book before me, and let the cows take care of themselves." This inquisitive nature—as well as the domestic instability of Germany, which descended into armed conflict the year he left—led him to seek his fortune in the New World.

Huning was entranced by the stories he heard upon landing in New Orleans of men like himself becoming wealthy in Cali-fornia and set out immediately for the West. With little money and no sponsorship, Huning got as far as Santa Fe. He spent a winter there, learning Spanish by clerking at a store and strug-gling his way through *Don Quixote*, the brilliant winter sun mak-ing the snow outside his window sparkle. Soon, he found work with a local trader, Gaspar Ortiz. Huning agreed to accompany Ortiz on the wagon trail to Chihuahua, but when they spent the night at a camp near Alburquerque, he met a fellow German named Simon Rosenstein, who was among the city's first Jewish residents and one of the only Anglos with a store on the plaza. "He persuaded me not to go to Old Mexico," Huning wrote in his memoir, "and offered me a situation in his store. I accepted."

Huning, improbably, saw opportunity in Alburquerque, whose first *R* was already beginning to fall out of use because many of the new Anglo arrivals struggled to roll their tongue with enough flair to pronounce it. Only around eight hundred people lived in town when Huning arrived in the early 1850s, many families having fled south during the war. The plaza in front of San Felipe de Neri church was quiet and harbored little beyond dust spiraling in the breeze, the wide sweep of space between squat adobe buildings only filled on the occasional market days, when farmers and craftspeople from the nearby pueblos set up shop. When the Rio Grande flooded—as it did almost every spring—huge puddles formed in the depressions of the dry earth. Migrating ducks would sometimes land in these temporary ponds, only taking flight after the merchants threw open their doors, shouldered their rifles, and fired.

After working for Rosenstein for a few years, Huning opened his own dry goods store on the plaza. At first, business was so bad that he ate practically nothing but beans and boiled eggs. Huning's fortunes rose as he made inroads with both the Hispano gentry and the U.S. Army, which had set up a garrison in Albuquerque. In 1863, Huning built a gristmill to process the wheat that was grown along the Rio Grande, as well as a sawmill for the timber being harvested in the Sandia and Jemez Mountains. By the 1870s, the bookish German had become one of the territory's leading merchants. With talk spreading that the Atchison, Topeka and Santa Fe Railway (AT&SF) was about to finally make good on its name and extend itself to New Mexico, Huning began buying land. By the time surveyors actually arrived, he controlled seven hundred acres south and east of the plaza.

Initially, the railroad was meant to cross the Rio Grande via a bridge in Bernalillo, but those plans were foiled by the unscrupulous José Leandro Perea. The Pereas had been one of the first Basque families to settle Bernalillo in the 1690s, and Don José—as he was widely known—was the acknowledged leading citizen of the town, owning much of its land and hosting General Stephen Kearney at his estate during the Mexican-American War. Now an old man, Perea was one of the few true million-aires in New Mexico. When officials from the AT&SF visited the grand sala of his estate, he scrutinized their maps, then asked how much they were willing to pay for the right to build on his land near the river. When they suggested that his holdings were worth no more than three dollars an acre, Perea recoiled. He was savvy enough to recognize the immense profit the AT&SF would reap as the first railroad in New Mexico—as well as the disruption it would cause to the freight wagon industry, which had helped build his fortune. Perea's response to the advance men was an eye-popping counteroffer: $425 an acre.

Without a word, the surveyors rolled up their maps, boarded the waiting stagecoach, and tried their luck in Albuquerque, where they found a more promising partner in Huning. He joined with two other Anglo businessmen to form the New Mexico Town Company, which began acquiring even more land east of Albu-querque proper before flipping everything to the railroad for just one dollar. In exchange, the three businessmen were entitled to half of whatever profit the railroad eventually garnered from the sale of any land it didn't end up building on. As for the Hispanos who controlled all the farmland around the plaza, they could only watch as the AT&SF started laying down the future economic heart of New Mexico a mile east of their town, which was already more than 170 years old.

The new railroad depot was completed in 1880. Within a few years, a commercial district of hotels, bars, and brothels sprouted up around the tracks on streets named Gold, Silver, and Cop-per, paralleling a brand-new Railroad Avenue, which connected New Albuquerque with the old Villa de Alburquerque. Two early arrivals were a pair of brothers driving a wagon down to El Paso in 1882, but who found themselves stranded near the new depot when one of their oxen died. Rather than continue on, they shrugged their shoulders and bought one of the dirt lots being sketched out by a surveyor for seventy-five bucks. The Strong brothers' general store sat at the corner of two dirt paths called 2nd Street and Copper; one block away, on Railroad, a massive saloon called the White Elephant soon opened. Word would eventually spread around the West of the White Elephant's solid mahogany bar, long enough to seat fifty men, as well as its gambling tables, where a traveler from Denver made $11,000 in a single night.

In 1883, Huning built a new home to solidify his position as Albuquerque's leading man. Known as the Huning Castle, the Italianate manor lorded over four hundred acres on the west end of Railroad Avenue, a marble fountain burbling next to a walkway of cement imported from England, surrounded by Persian walnut trees and Lombardy poplars. The fenced-in grounds hosted an aviary, a block-long grape arbor, and a pond where a flock of white geese paddled about. The centerpiece of the home was a three-story tower, meant to recall the keep of the German castles Huning remembered from his youth. Despite his important role in New Albuquerque's founding, the castle became a private retreat as Huning got older, his wiry beard growing thicker, and his trips outside the walls dwindling. On the fence that faced New Albuquerque, he posted a sign making clear his refusal to be bothered by anyone at all: "Trespassers will be shot."

With the likes of Huning retiring from public life, it fell upon a new generation of Anglo New Mexicans to push for their territory's belated admission into the Union as a state. Though New Mexico had sufficient population to be granted statehood as soon as it was signed over to Washington, 90 percent of the sixty thousand people who lived there were neither white nor Black nor Indian, which scrambled the simplified racial calculus that undergirded the American legal system. Indeed, Mexico's mestizaje tradition may have even saved the country from being entirely conquered by the United States, as even the most imperial minded of legislators doubted that its population could be absorbed into a white supremacist nation. "Ours is the government of the white man," John C. Calhoun, a vehement defender of chattel slavery, declared on the floor of the Senate as the war was winding down. "The great misfortunate of what was formerly Spanish America is to be traced to the fatal error of placing the colored race on an equality with the white. That error destroyed the social arrangement which formed the basis of their society. This error we have wholly escaped."

Though Mexicans were supposed to be granted full citizenship by the Treaty of Guadalupe Hidalgo, that status, as well as their right to vote, was not confirmed by the judiciary until 1897. Pueblo Indians had it even worse. Their voting rights—likewise enshrined by the treaty—were gradually attenuated until 1876, when their citizenship itself was invalidated by the Supreme Court. Puebloans would not officially regain the franchise until passage of the Indian Citizenship Act, nearly fifty years later. Even then, most Indians in New Mexico were prevented from voting until 1948, when Miguel Trujillo, a member of the Isleta Pueblo who had fought as a Marine in World War II, filed a legal challenge to a provision of the state constitution that prevented "Indians not taxed" from voting, making it impossible for anyone living on a reservation to register.

In the 1870s, New Mexico's Anglo boosters embraced a novel workaround to the broad refusal to recognize Nuevomexicanos as fully American: casting them not as Mexicans, but as the pure-blooded descendants of intrepid Spanish conquistadors. Those claims proved unpersuasive in Reconstruction Washington, particularly among southern Democrats who resented that New Mexico had largely sided with the Union during the Civil War. In 1876, a congressional report dismissed the argument that Nuevomexicanos were Spanish, writing, "Of the native population but few are pure-blood or Castilian, probably not more than fifty or one hundred families in all, the rest being a mixture of Spanish or Mexican and Indian in different degrees." When another statehood bill came up in 1888, legislators were even more forthright in their dismissal, with one report calling the territory's inhabitants, "ignorant and degraded," and making a special point to label Santa Fe—whose name was then synonymous not with art-collecting retirees but with the corruption of the land-speculating Santa Fe Ring—"one of the most reckless and miserable towns on the globe." In 1893, the *Chicago Tribune* ran an editorial arguing that New Mexicans were "not American, but 'Greaser,' persons ignorant of our laws, manners, customs, languages, and institutions."

Heedless of the perceptions of Anglos elsewhere, many of the people who had once called themselves Mexican gladly embraced the idea that they were actually Spanish. An editorial in the newspaper *La voz del pueblo* defended the patriotism of "those in whose veins runs the blood of Cortez, Pizarro, and Alvarado." In 1901, a lawyer in the city of Las Vegas (New Mexico, that is—the municipality in Nevada had not yet been founded) gave a speech proclaiming, "No other blood circulates through my veins but that which was brought by Don Juan de Oñate."

Politicos in Washington remained unconvinced, as did the rising class of journalists so infatuated with the sense of possibility engendered by the blank spaces on the newly redrawn map of America. Chief among those writers was William Smythe, who spent a decade early in his career promoting agricultural development in the Southwest. His hopes for Washington to prioritize folding the lands won from Mexico into the American mainstream were dashed by the outbreak of the Spanish-American War in 1898, in which the United States followed the playbook it had pioneered fifty years earlier: manufacturing a conflict to justify seizing and colonizing Puerto Rico, Cuba, the Philippines, and Guam.

In *The Conquest of Arid America*, published in 1900, Smythe argued that the focus of American settlement should instead be on the North American interior. "The advocates of colonial expansion abroad argue that hitherto we have been engaged in the conquest of this continent, and declare that this work is now done," he wrote. "But it is *not* done." To secure the nation's imperial reach—not just in the Pacific and Caribbean, but also by colonizing Alaska and digging the Panama Canal—Smythe believed it was necessary to first develop the land taken from Mexico. Though he acknowledged the presence of Indigenous and Mexican inhabitants of the Southwest, he did so only as a way to frame the need for settlement. Even if the Mexicans around El Paso had "made a beautiful agricultural and horticultural district" along the banks of the Rio Grande, he still insisted that "their methods are crude and ancient." By the same token, he wrote of New Mexico, "The growth of the white population has been slow, but will increase rapidly with the development of irrigation." Evidently, land—when watered—would sprout white people.

A few years earlier, the journalist Charles Fletcher Lummis

had evoked the Southwest's bivalent character in his fanciful book on New Mexico, *The Land of Poco Tiempo*, though he sought to preserve the region's original civilizations rather than speed their decline. Lummis admitted that a "few semi-bustling American towns wart the Territorial map" but stated that "with them I have nothing here to do." Instead, he exoticized the "picturesque" people and landscapes of New Mexico, writing,

> It is a land of quaint, swart faces, of Oriental dress and unspelled speech; a land where distance is lost, and the eye is a liar; a land of ineffable lights and sudden shadows; of polytheism and superstition, where the rattlesnake is a demigod, and the cigarette a means of grace, and where Christians mingle and crucify themselves—the heart of Africa beating against the ribs of the Rockies.

Populating this strange land that was both Asian and African—but not, apparently, American—were "the real autochthones, a quaint ethnologic trio." That included the Pueblo Indians ("peaceful, fixed, house-dwelling and home-loving tillers of the soil"), the Navajo ("sullen, nomad, horse-loving, horse-stealing, horse-living vagrants of the saddle"), and, "last of all, the Mexicans; in-bred and isolation-shrunken descendants of the Castilian world-finders; living as much against the house as in it." Lummis resented how the new railroads disrupted the ability of visitors like him to make sport of observing New Mexico's original residents as if they were performers in a grand burlesque. "The present is a husk—the past was a romance and a glory."

Despite their ideological differences, both Smythe and Lummis endorsed a view of the Southwest as a place torn between Anglo industry and the Indian and Hispanic picturesque. Lummis disdained the modern and longed for a vanished era, while Smythe cordoned non-Anglos into their own districts where they could be either safely ignored by settlers or engaged "as useful laborers in the simpler agricultural tasks." What, then, would become of Albuquerque? It was dandy to be picturesque, but that was hardly compatible with the efforts of speculators to lure the investments from back East that would allow them to turn the land they had seized into capital.

New Albuquerque was formally incorporated in 1891 with a population of 3,785, surpassing the 1,700 residents of what was now called Old Albuquerque. Emboldened, the Anglos who had already made a fortune kept pushing growth, and their budding city quickly spread out to meet the new territorial university sited on the sand hills overlooking the railroad tracks from the east. New Town also absorbed the historic village of Barelas (whose proximity to the south made it ideal for the AT&SF's machine shops) and gobbled up Franz Huning's newfangled Highland Addition to the northwest.

Even as Anglos touted New Town as an inchoate commercial and industrial powerhouse entirely divorced from the Spanish and Mexican past embodied by Old Town, they recognized that the Indians, at least, might prove a novel draw for eastern tourists who had begun gravitating to the Southwest once the railroad made it possible to travel there comfortably. Around the turn of the twentieth century, the AT&SF began paying painters from the newly established art colonies in Santa Fe and Taos to create stereotypical portraits of Indians that could be used to advertise the railroad, renaming its trains the *Super Chief* and the *Navajo*.

In 1926, the AT&SF began offering its riders an "Indian Detour" in conjunction with the Fred Harvey Company, which operated the railroad's legendary dining cars and hotels. These detours lasted three days, during which guests would be driven from Albuquerque to nearby pueblos to gawk at the peoples they had previously only read about. "The America of Coronado waits for you beside this motor trail," read one print ad, beneath a portrait of a dark-skinned Navajo riding a horse, his lap covered by a blanket,

> an enchanted land, where for three days your luxurious Harveycar carries you on a personally-escorted tour of ancient Indian pueblos and prehistoric cliff-dwellings in

the New Mexico Rockies between Las Vegas, Santa Fe, and Albuquerque.

Ramming Coronado and the ancestral Puebloan's cliff dwellings together in one sweeping image, the AT&SF sought to fuse Spanish and Indigenous history into an idyllic past that guests could glimpse in the backcountry of New Mexico for the equivalent of more than $1,000 per person. Wealthy travelers flocked to the Indian Detour, with Albert Einstein, Eleanor Roosevelt, and Harry Guggenheim all booking a trip. The newfound access to native communities was a particular draw to the first generation of filmmakers. In their silent movie *Primitive Indians of the Painted Desert*, William and George Allen exoticized the Hopi along the same lines as Charles Fletcher Lummis. They filmed a Hopi man planting corn in what one title card called the tribe's "terraced gardens like those of the Japanese"; later, a woman weaves the type of willow basket "found only in one other part of the world—in Africa on the upper Nile."

Locals at the depot in Albuquerque capitalized on tourist interest in New Mexico's tribes, hiring merchants to spread blankets outside the platforms from which they could sell jewelry, textiles, and pottery. After the AT&SF's Hotel Alvarado was completed in 1902, its operators designated one section of the massive complex as the "Indian Building." Inside, visitors could browse goods sourced from tribes all over the Southwest and watch Pueblo weavers at work.

The only compensation the Pueblo, Navajo, Apache, and Hopi peoples who were treated as safari subjects by the AT&SF were entitled to was whatever money they could wrest from the tourists for the pottery and blankets they began specifically making to sell. Likewise, the Indians who were lured to New Albuquerque to take part in the spectacle at the Alvarado were never invited to have any say in the city's future. Rather than treat Native Americans as stakeholders in the growing community, Anglos assigned them the role of tourist attraction that might provoke a frisson of excitement in the hearts of newcomers reared on the legendary exploits of Kit Carson, Wyatt Earp, and Wild Bill Hickok.

On the few occasions when Indians dared to demonstrate their own agency, retaliation was swift and brutal. When a territorial fair was held in 1903, a real estate tycoon and future mayor of the city known as Colonel D. K. B. Sellers—who never actually served in the military—invited a party of Navajo to perform in a pretend battle with some authentic members of the U.S. Cavalry. According to Sellers's script, he would play the lonesome wanderer, riding a horse into the center of the fairgrounds, where he'd be set upon by a band of Diné horsemen led by a man whose name was recorded as Peshi Cli. Some cowboys would swoop in to save Sellers, but they would find themselves overmatched until the Cavalry arrived, sending the Navajo packing.

Erna Fergusson, a granddaughter of Franz Huning who became one of New Town's original historians, later recalled that it was "a good clear day . . . Indians were ready with Peshi Cli in command. Cowboys resplendent beyond belief. Cavalry rested in the shade while their mounts cropped grass growing among the pigweed along the track." Suddenly, Sellers's translator noticed that Peshi Cli and his men had real guns strapped to their belts, loaded with real ammunition. Peshi Cli and his men were arrested and thrown in jail under suspicion of using a make-believe battle as a pretext to exert retribution for a recent confrontation between the Navajo and one of the Cavalry's captains.

Anglo authorities were scandalized by the prospect of the Diné carrying real weapons. Never mind that the tribe was less than fifty years removed from the Long Walk. Such generational grievances were ignored by the Anglos who had taken over Albuquerque. Indians were too often treated as little more than a useful prop, a breathing demonstration of the quirky charm New Mexico brought to the American experiment. To that end, Anglo leaders even sought to strategically deploy Native Americans in their ongoing campaign for statehood. When the newspaper baron William Randolph Hearst paid for a delegation of congressmen to visit the territory in 1906, Colonel Sellers hired another Navajo, a man called Hosteen, to help him greet the dignitaries at the train station when they arrived in Albuquerque. Sellers had Hosteen stand next to him and, through a translator, instructed him to speak in Diné whenever Sellers pressed on his foot. Once Hearst's party had alighted on the platform from their railcar, Sellers stomped on Hosteen's toes and, once he was done speaking, explained to the legislators that the man was expressing not only his desire for statehood but also his personal gratitude to the federal government for having civilized his people.

Despite the efforts of Albuquerque's boosters to use New Mexico's history to draw in tourists, it was the city's climate that ultimately precipitated its expansion. In the early years of the twentieth century, tuberculosis patients began moving to the Southwest because of the desert climate's supposed restorative effects. One estimate has it that, by 1915, one out of every five residents of New Town was a so-called lunger, and their numbers only grew from there.

The crowd of lungers who arrived in the 1910s included my Chicago-born great-grandfather, Nathan Glassman—his mother, Dora, suffered from tuberculosis. Since money was tight, Nathan and his brothers alternated working and attending Albuquerque High School. Nathan was handsome but taciturn, the kind of man who spoke only when necessary, preferring actions over words. That translated into unusual skill as an athlete: though only five foot four, Nathan played halfback on Albuquerque High's football team, where he was remembered for zipping around linemen "as if he were Mercury upon the wings of wind."

After getting an accounting degree at UNM, Nathan met Ethel Backel through their shared synagogue, Congregation B'nai Israel, and they settled down in a freshly platted subdivision, right around the corner from Monte Vista Elementary School. At the time, the Monte Vista Addition was the far eastern extent of the city; between the handful of homes that had been built there and the towering Sandia Mountains extended miles of dry mesa riven by natural arroyos that channeled winter runoff and monsoon rains toward the Rio Grande.

My grandmother, Sandra, was born in 1935. She grew up as the Monte Vista Addition and the neighboring developments of Granada Heights and College View filled in. Meanwhile, Old Town was in decline. Although it was still thriving well into the 1920s—the municipality's fortunes bolstered, in part, by its indifference to the speakeasies and backroom gambling halls that proliferated there after New Mexico outlawed the sale of alcohol in 1917—the double whammy of the Great Depression and a decade-long drought upended the ability of many residents to maintain the farms their families had operated for generations. Once World War II began, many of the longtime residents of Old Town and the agricultural land that surrounded it left New Mexico entirely, heading to Los Angeles and Denver in search of work.

For the Anglos of New Albuquerque, the war had a totally different effect. The newly established Kirtland Air Force Base soon grew to be one of the city's largest employers, helping the city to triple in size by 1950, to a population of nearly a hundred thousand. The acequias that had fed farms near Plaza Vieja for centuries were filled in, the land itself sold to developers. Though the Hispanos who did remain resisted annexation for several years, in 1947 Old Town was finally swept into New Albuquerque, finalizing the union of the two cities. Once they had the plaza under their control, the city's Anglo elite set about rebranding it as a tourist attraction, one that might lure motorists traveling Route 66. To complete the image of Albuquerque as a redoubt of ancestral Spanish culture (never mind its contemporary composition), they ignored the fact that the city that had engulfed the original town had been founded in living memory and instead adopted the settlement of the Villa de Alburquerque in 1706 as the city's official birthday.

Only once Old Town had been assimilated into the American metropolis did Anglos become willing to promote Albuquerque's Spanish origins with the same benighted enthusiasm they had once showered on the pueblos. The boosters who oversaw the transformation of the neighborhood around the plaza from an agrarian village to a cutesy quarter of tacky stores and overpriced restaurants styled themselves not as opportunists, but as saviors. Bette Castee, an Anglo woman who was instrumental in the process, would later write that, though many saw the "adobes crumbed into dust and mud" as a sign that Old Town was "rundown and useless," proponents like her were "visionaries who saw something more, and set out to prove the area's worth."

Albuquerque's first neon sign was put up to mark the route

to Plaza Vieja in 1949, and over the next fifteen years dozens of quirky shops sprouted up to sell handicrafts and trinkets to out-of-towners passing through on their way to California. Though the city was glad for all the visitors who came to stay at the glitzy new motels lining Central—a renamed Railroad Avenue—the cultural practices that had previously thrived in the plaza went nearly extinct: livestock were banned within Albuquerque city limits, ruining the livelihood of what few farmers did remain, and Hispanic church groups had to get special permits for the Sunday dances they had been holding for decades.

Over on the East Mesa, these struggles hardly registered. By 1950, Albuquerque had become a genuine city, far surpassing Santa Fe as the economic heart of New Mexico. Like cities all over the nation, the postwar years saw Albuquerque shun its historic core as it promoted generic suburbanization, seeking to attract new residents by making them believe that the only difference between the Duke City and Cleveland or St. Louis was the weather. It worked. Albuquerque's limits expanded for miles in every direction, jumping the Arroyo Embudo, nibbling away at the irrigated land of Los Griegos and Los Ranchos, and butting up against the edge of the Air Force base. To the north and south, only the pueblos of Sandia and Isleta remained, the sole survivors of a four-century project to transform the middle Rio Grande Valley from the verdant province of Tiguex into a hub of continent-spanning colonial enterprise.

Darkness was starting to fall when I got off I-40 in Holbrook, speeding over the red-rock flats beyond the Petrified Forest toward the Mogollon Rim. I'd spent the day in Window Rock, killing time at the Blake's Lotaburger just over the New Mexico line before the Navajo Nation Council assembled. Sitting in the window, I munched curly fries and watched the clouds roll through the too-blue sky over the backside of Tségháhoodzání until it was time to go to the Council Chamber and take notes as delegates from the Naabik'íyáti' Committee presented on some of the issues facing the tribe: cleaning up old uranium mines, preventing sex trafficking. As the agenda shifted toward procedural matters, my attention strayed to the octagonal walls of the chamber, which were illustrated with a 1943 mural by Gerald Nailor that sought to capture the full sweep of Navajo history. From west to east, the mural moved from depicting Diné hunting and gathering in Canyon de Chelly, to first contact with silver-helmeted Spaniards and blue-coated Americans, to the original sheep herds, to the signing of the peace agreement that ended the Long Walk, to the forced livestock reductions of the '30s. Nailor's style was flat but expressive, with each figure in the mural tell-

ing its own story—even a sheep bleating in fury as a bemused shepherd tried to hustle it into a pen.

The semitrucks had formed a steady procession on I-40, but I had State Road 337 all to myself as I cut south across the pink sand, the only feature in the distance a towering power line. Off to the east, I caught a glimmer of white from a wind turbine—part of the Dry Lake project, the first commercial-scale wind farm in Arizona. As the road began its subtle rise and scrubby bushes and thin grass were reintroduced into the landscape, the sun met the earth and the sky turned orange, with a vivid pink stripe at the horizon that got brighter and brighter before seeming to be extinguished in an instant.

I stopped for gas at the intersection in Heber. Only once I was outside did I realize that there were trees all around, the smoky-sweet smell of piñon and juniper prompting a sense memory of the Sandias. As I kept driving, I felt the trees drawing closer in the darkness all the way to Payson. After I passed the place where they serve spaghetti "by the bucket," I pulled up at the traffic light to turn onto 87. In the lane next to me, there was a truck with flagpoles mounted in its bed, holding banners reading "Don't Tread on Me" and "Make America Great Again."

After passing by Mazatzal Peak, the earth gave way. Signs on the median shifted from warnings about ice on the road to specification of the downhill grade and an arrow indicating a runaway truck ramp. It felt like I was slaloming out of one world and into another. Once 87 finally straightened out, my headlights began picking up saguaros. I had expected at some point on the incline that I might find a vista of the Salt River Valley that would allow me to glimpse the full expanse of metro Phoenix in one swoop. This was not to be. There were a few lights glimmering in the foothills, sure; a pale yellow tint to the night sky. But all of a sudden I was crossing the Salt and merging onto one of the ten lanes of the 202 loop, racing toward downtown Phoenix without quite realizing I'd arrived anywhere at all.

DESTINY

People talk about environment.
We're doing something about it.

—FLOYD DOMINY

When the offices of *Arizona Highways* were built in the 1950s, Phoenix was entirely contained between the Salt River and the granite hump of Camelback Mountain. Though 175 years younger than Albuquerque, Phoenix was already the larger city, having just surpassed a population of one hundred thousand. The magazine's new offices were at the fringe of the rapidly expanding city, near grain elevators, the state fairgrounds, and a golf course. Nearby ran a spur of the Atchison, Topeka and Santa Fe Railway that extended 140 miles north to Flagstaff, where it connected with the route between northern New Mexico and Los Angeles. Back in the '50s, the banks of the channels that irrigated the Valley of the Sun's plentiful farms with water from the Salt and Verde Rivers were still shaded by cottonwoods, tamarisks, and willows; kids played on swings hung from some of the larger branches while their parents threw lines into the water rigged up with lures for bass and catfish. Up in Scottsdale, daring teenagers water-skied along the canal that ran east to the Verde, clutching onto a rope tethered to a friend's truck as it trundled along the dirt road that mirrored the channel of water. Just seventy years later, that low-key version of the Valley of the Sun now seems borderline mythological.

In the twenty-first century, Phoenix became the fifth-largest city in the United States. Over the course of a human lifetime, the agricultural hub has transformed into a metropolis of 5 million people, its footprint exploding by a factor of 265 as it spread across more than 1,200 square miles. Once distant, the Fountain Hills to the northeast and Estrella Mountains in the southwest now squeeze the city's highways, subdivisions, and shopping centers into a diagonal alignment, squirting out at the far corners of Buckeye and Apache Junction into open, unreclaimed desert.

All that growth has made Phoenix the most extreme urban heat island in the country, a place where temperatures at the center of the valley sometimes clock in 15 degrees Fahrenheit higher than in the desert beyond. Between all the asphalt and the regional temperature shifts driven by climate change, summer highs in the city jumped 9 degrees between 1950 and 2020, turning the already brutal summers into a dire sauna that produces hundreds of casualties every year. The city's canals, which in an earlier era provided relief from the naturally sweltering conditions of the Sonoran Desert, had long since been stripped bare of any organic material and closed to the public as they were efficiently reengineered into open-air water mains meant to serve cul-de-sacs instead of cotton farms. The area around *Arizona Highways*, meanwhile, was now occupied by warehouses, self-storage lots, and offices for the state's departments of public safety and transportation, the latter of which owned and operated the magazine.

I was visiting *Arizona Highways* in hopes of answering a question that has bothered me for most of my life: How did Phoenix get so much larger than the cities whose existence predated it, sometimes by more than a century? As a teenager, my friends and I mocked Phoenix. Albuquerque made sense. We had the Rio Grande. They had the unromantic Salt. We had a history of continuous human civilization that dated back millennia. They barely had a hundred years. We had summer heat. They had summer death. Plus, our city boasted real seasons: cottonwood leaves changing from green to yellow in the fall; snow gusting over the foothills in the winter; chollas flowering pink in the spring. What did Phoenix have, aside from the same glowering sun, day after day after day?

The building's entrance is guarded by a baroque set of double doors depicting a thunderbird with inlaid wood. I grasped the bird's claws and pulled. Inside, the offices of *Highways* were like that of most magazines—the layout for the current issue was tacked against the wall of a central corridor, and every desk was stacked with proofs, photo prints, and dog-eared books. There were a few key differences, though, most notably the gift shop the thunderbird doors opened onto, where visitors were invited to peruse the same *Highways*-branded calendars, dish towels, and Christmas ornaments that were on sale at the magazine's store in Terminal 4 of the Phoenix Sky Harbor.

This proliferation of branded kitsch struck me as unusual for such a narrowly focused publication, but as editor Robert Stieve explained to me, "We have this readership, they feel like they own the magazine. They feel like they're part of it, and they take pride in it."

Much of that pride stems from how synonymous *Arizona Highways* became with the postwar boom that transformed Phoenix into the preeminent city of the Southwest. When I spoke to Win Holden, a former publisher of *Highways* who grew up in Chicago, he recalled noticing an issue on the coffee table of his grandmother's house in the 1950s. The cover featured a saguaro. "I made a typical five-year-old comment, like, 'C'mon, there's no tree like that!' And instead of telling me I was a moron and to go to my room, she said, 'Well, let me show you.'" They went through the magazine page by page; Holden was enthralled. So were the other two hundred thousand subscribers to the magazine at the time, 93 percent of them living outside Arizona, *Highways* their only reference point for the faraway landscape of arid America.

Plenty of those mid-century subscribers, like Holden, were eventually drawn to Arizona, and most of those newcomers ended up in Phoenix, the city whose freshly paved subdivisions were touted by *Highways* as offering "as much as thirty-percent more house and luxury features" for the same money that could be spent elsewhere. Though *Highways* devoted less attention to real estate developments and the public works projects that made them possible than it did to immersive photo spreads of Arizona's natural wonders, each component of the magazine worked in harmony to overhaul the image of the state from a wasteland to a playground.

The man who built *Highways* into an institution was Raymond Carlson. A former speechwriter and World War II vet who took charge of *Highways* when it was still geared toward transportation engineers, Carlson was as ambitious as he was shrewd; a perfect match for a droughty backwater whose name suggested presumption more than a rise to glory. With his blond hair and broad smile, Carlson was a natural fit for the all-American coterie of charismatic boosters who had likewise arrived in Phoenix in the early years of the twentieth century and set about molding it into the city of the American future.

"Everything we do is trying to hit the bar that Raymond Carlson created back in the golden days in the '40s and '50s," Robert Stieve told me. Deeply tanned and with a few days' growth of gray stubble, Stieve looked as if he'd come directly to the office from a rafting trip on the Colorado. Burbling with energy, he interrupted our conversation several times to rummage through the drawers, cabinets, and bookcases of his office looking for ephemera Carlson had left behind—a yellowed letter, a newspaper clipping, a snapshot—and offered each as evidence of a single man's effect on an entire state's image. He told me about Carlson's friendships with Ansel Adams and Barry Goldwater, both of whom regularly contributed photographs to the magazine, as well as Frank Lloyd Wright, who designed the editor's turquoise-accented home on West Palo Verde Drive.

Then Stieve pivoted in his chair to find a copy of an editor's note Carlson had written in 1963, which he planned to excerpt for an upcoming anniversary issue. "Breathtakingly beautiful land, a big land full of sun and distance, so complex in personality no person will ever know all of it," Stieve read aloud, parroting his predecessor. "That is the Arizona story, so formidable and awesome."

. . .

Though the city of Phoenix was only founded in 1881, the stretch of the Salt River that it now occupies is in fact one of the oldest agricultural regions in North America. The ancestral Sonoran Desert peoples first began making the transition from hunting and gathering to sedentary agriculture around 2000 B.C.E. This was the same era during which, around the world, the Epic of Gilgamesh was being formalized into written Sumerian during the Third Dynasty of Ur, Yu the Great was taking over the domain of Xia after stemming the flooding of the Yellow River, and Mentuhotep II reunified Egypt. Civilization, as we know it, was born in the arid lands. With time, the banks of the Nile, Yellow, Tigris, and Euphrates filled with farms and a thriving population, and the same process repeated itself in ancient Arizona.

It was around 300 B.C.E. when the ancestral peoples of the Sonoran Desert are believed to have started planting corn in the valley, after the crop was brought north from Mesoamerica. Urbanization was the inevitable result of their embrace of a crop that provided such reliable sustenance: as more and more people came together to stabilize the food supply, communities grew large enough to eventually require administration, however informal. The first settlements were established along the Salt and Gila Rivers, before expanding to the Santa Cruz.

By 450 C.E., the culture that archeologists call Hohokam had emerged across the river valleys of the Sonoran Desert (the name is a corruption of an O'odham word that is often translated as "ancestor"). Small villages were formed by a handful of families who built their shelters in close proximity to each other, usually pit houses protected from the sun by roofs made from saguaro and ocotillo ribs coated in wattle and daub. In between those shelters were ramadas, free-standing sunshades under which villagers prepared food and made pottery. The early Arizonans also constructed eye-shaped courts nearly the size of football fields, where they played a ball game originating in Mesoamerica. All of this was made possible by the ancestral Sonoran Desert peoples' agricultural prowess. Rather than growing individual crops in furrowed rows, they grouped corn, squash, cotton, and beans together in a succession of earthen mounds, a method of companion planting that ensured the soil's continuing fertility.

Forty thousand people lived in villages along the Santa Cruz, Salt, and Gila Rivers in the early 1300s, when their population began a precipitous decline. By then, the instability of the climate had driven many ancestral Puebloans into the Hohokam lands from their cliff dwellings on the Colorado Plateau. Outlying Hohokam villages like those in today's Tucson Basin were abandoned, and the population became concentrated in the larger settlements along the Salt and the Gila. Birthrates fell, and these villages gradually shrank until their occupants could no longer maintain their canals.

When Anglo settlers arrived in the newly designated Arizona Territory in 1863, it had been hundreds of years since the Salt River had hosted a permanent settlement. The entrepreneurial John Y. T. Smith—the Y. T. stood for "Yours Truly"—was one of the first homesteaders to set up shop near the river. Smith had enlisted with the California Volunteers at the outbreak of the Civil War and was stationed at Yuma; afterward, he received a commission to oversee a civilian wagon train to the distant Army garrison of Fort McDowell, two hundred miles away in the lee of the Mogollon Rim.

As the lead rider, Smith would guide the wagons in parallel with the Gila River as it cut through the basin-and-range country of the Sonoran Desert, where an unending succession of largely barren mountains presided over plains of yellow sand, and saguaros were the only vegetation that rose above the height of a saddle. Where the Gila met the Salt, at the foot of the modest Estrella Mountains, Smith followed the Salt further east, only turning north for Fort McDowell once the Verde River intruded, just past the first peaks of the Usery Mountains. On one of these trips, Smith noticed that a great deal of wild hay was growing along the banks of the Salt. After his commission ended, he decided to settle in the neighborhood, harvesting what grew naturally and hauling it fifteen miles north to sell at Fort McDowell.

Not long after setting up this modest business, Smith is supposed to have taken a trip to the gold rush boomtown of Wickenburg. After hitching his horse and ambling into a saloon, Smith struck up a conversation with a Confederate veteran known as "Tragic Jack" Swilling, who had opportunistically set up a stamp mill in Wickenburg as a way to profit off the prospectors. In the only portrait that survives, Swilling is seated with his knees spread wide apart and a pistol resting on his right shoulder, a cowboy hat in his opposite hand and an overgrown mustache on his lip.

Swilling spent most of his adult life as a soldier and roustabout. He was seriously injured in the 1850s when an assailant fractured his skull with a blow from a revolver, then shot him in the side. The bullet was never removed. "No one knows what I have suffered from the wounds," Swilling once wrote, "at times they render me almost crazy." In addition to being a heavy drinker, he self-medicated with morphine, as "the craving of it was greater than my will could resist." Despite the haze of whatever Swilling had been ingesting that night in Wickenburg, he was intrigued by Smith's description of ancient irrigation canals where wild wheat grew.

Quickly, Swilling hit upon a scheme for repurposing the ingenuity of some vanished Indians to the white man's advantage. He managed to snare a handful of investors for his plan, starting with Smith and an itinerant English nobleman named Darrel Duppa who claimed that he had attended Cambridge and could read ancient Greek and Latin, as well as quote Shakespeare from memory. After hiring a dozen hard-luck types from Wickenburg, the Swilling Irrigation and Canal Company completed its first channel in 1868.

After the first successful harvest, the twenty people involved in Swilling's company gathered to name their new settlement, which was being laid in the area around today's 28th Street and Van Buren. All agreed that the tentative moniker, Pumpkinville, wouldn't do. Several participants favored "Salina," after the river, while Swilling offered "Stonewall," in homage to Stonewall Jackson. After much discussion, the group landed on Phoenix, a nod to the lost civilization that had previously thrived on the spot and that the men now imagined they would restore.

Unremarked upon was the fact that the Akimel O'odham, one of the tribes descended from that supposedly vanished people, were still farming the banks of the Gila River, just twenty miles south. When U.S. Army Lieutenant Nathaniel Michler visited the Akimel O'odham in 1855, shortly after the Gadsden Purchase assigned their lands to the United States, he was amazed by their farms, which were more sophisticated "than anything we had seen since leaving the Atlantic States." Though the banks of the Gila were generally sandy and alkaline, where "little grass grows excepting in spots subject to overflow," when Michler's party reached an Akimel O'odham village they found land "irrigated by many miles of acequias, and our eyes were gladdened with the sight of rich fields of wheat ripening for the harvest . . . they grow cotton, sugar, peas, wheat, and corn."

A thriving Indigenous agricultural region being so near the new settlement of Phoenix was incompatible with the messianic visions of its founders. Instead of learning from their neighbors, they followed the popular gospel that "rainfall follows the plow," an assurance that even the driest stretches of North America could be turned temperate by sufficient cultivation. Settlers like Swilling didn't preoccupy themselves with wondering why the people who had built all the canals they were now reviving weren't around anymore; they just seeded acre after acre of wheat. The early returns were promising. The late 1860s and '70s were an unusually wet spell for the region, a trend that seemed to validate the belief that the arrival of Anglo horticulture had changed the climate for good. In just three years, Phoenix expanded to a community of three hundred people, with twelve hundred acres planted.

In short order, Phoenix had become the only settlement in the Southwest where the widely held notion that the desert was a blank slate for development was somewhat true. In 1891, a newly formed chamber of commerce boasted, "Here are none of the sleepy, semi-Mexican features of the more ancient towns of the Southwest." In *The Conquest of Arid America*, William Smythe wholeheartedly endorsed the new city, writing,

The Salt River Valley is the glory of Arizona. Approaching it from either of the transcontinental railways the traveler sees naught but the gray desert soil, marked by

the gnarled branches of the mesquite and the slender pillar of the cactus. Even the mountain-sides appear to be
devoid of verdure and tanned to a dark brown by the sunshine of centuries. But suddenly all the beauties of the
Garden of Eden burst upon the astonished gaze of the
visitor. Wherever the waters of irrigation have moistened
the desert, and man has planted the seed of grass, flower,
or tree, the most luxuriant vegetation has sprung from
the soil to revolutionize the appearance of the country.

Even as Smythe's description of Phoenix echoed what
Nathaniel Michler had written about the Akimel O'odham's villages a few decades earlier, the journalist insisted that the new
city had "risen from the ashes of a forgotten people," and served
as "the pulsing heart of the new life of Arizona. Here are modern
business blocks, handsome public buildings, busy stores, a promising university, and hundreds of beautiful homes resting under
the shade of palm, magnolia, and pepper-trees." While Smythe
allowed that, across Arizona, "there are many of the lower class of
Mexicans, they are much less numerous here than in New Mexico, and much more widely diffused over the Territory."

As the historian Bradford Luckingham later observed, "Phoenix did not exist through Spanish and Mexican periods. It was
founded largely by Anglos, for Anglos, and they were determined
to transplant their familiar cultural patterns to their new home."
Early Phoenicians quickly recognized that their settlement's lack
of history could become a selling point to Anglos interested in
moving out west. When the city's chamber of commerce published a directory in 1920, it described the city as "a modern town
of forty thousand people and the best kind of people, too. A very
small percentage of Mexicans, negroes or foreigners."

The only real barrier to the city's future growth that the
boosters entertained was the wild fluctuations of the arid climate.
The plentiful rains of the 1880s led to the Salt River's flooding in
1890, and then again a year later, knocking out a railroad bridge
that had just been completed between Phoenix and Tempe and
causing the adobe buildings south of Washington Street to deli-
quesce back into mud. That summer brought a paltry monsoon

season, so the farms that had survived the flooding found themselves suddenly starved for moisture. As if a switch had flipped,
a drought began that lasted for more than a decade. The Salt
River was reduced to a trickle, many of the canals went dry, and
thousands of farmers who had recently arrived packed up again
and left, either returning to the East or heading for the Pacific.
Indigenous farmers in the Southwest had long contended with
the inconvenient habit of desert rivers to flood whenever they
weren't evaporating away, but while many tribes practiced a more
flexible style of agriculture that allowed for shelters to be moved
and new canals to be dug, Anglo settlers were committed to the
lines drawn on paper that divided their property from that of
their neighbors. Moving was not an option. For Phoenix to survive, the river had to be tamed.

It was out of this ordeal that the old conviction that rain would
follow the plow was replaced by a new one: reclamation. Luckily,
aspiring city builders across the Southwest had a champion in
Theodore Roosevelt. Shortly after he assumed the presidency in
1901, Roosevelt gave a speech stating his belief that "the western half of the United States would sustain a population greater
than that of our whole country today if the waters that now run
to waste were saved and used for irrigation." A year later, he
signed the Bureau of Reclamation into law. The agency's first
dam was situated on the Salt River seventy-five miles northeast
of Phoenix, deep in the Mazatzal Mountains. After the structure
was completed in 1911, it was named for Roosevelt himself, who
boasted that after the Panama Canal it was the most significant
achievement of his presidency: not only would it ensure a reliable
supply of water for the farmers who made up the Salt River Valley Water Users' Association (soon to be renamed the Salt River
Project), it also set the mold for the reclamation of every other
corner of the arid West.

The name of the Bureau of Reclamation was strongly
informed by the Manifest Destiny era whose promise it was
founded to fulfill. Effectively, the agency's mission was to build
dams and reservoirs to provide water and power to the army of
aspiring aggies who began claiming homesteads in the nation's
driest corners after 1877, when the amount of land settlers could

purchase from the federal government was doubled—so long as it was designated as arid. Instead of stating that mission plainly, the bureau instead promulgated the idea that the desert was, in some theoretical past, fertile farmland that the federal government meant only to reclaim. By the same token, the grandiose name disguised the simplicity of the agency's method: eliminating the uncertainty over water resources caused by the natural variation between wet and dry years by storing extra runoff from the winters with bountiful snowpack in reservoirs, which could be tapped in leaner times to guarantee farmers a reliable supply of water to grow their crops, no matter the climate.

As the policy of reclamation was gaining supporters across the Southwest, it was impossible to avoid the question of what would become of the region's mightiest river: the Colorado. From its headwaters in the Rocky Mountains, the Colorado charts a 1,450-mile path through the desert, before emptying into the Sea of Cortez, which separates Baja California from the rest of Mexico. Over a span of millions of years, the Colorado carved through the Southwest's crimson limestone and shale, creating the canyons we now call Ruby, Cataract, Marble, and Grand.

Given that California, then as now, had a larger population than the rest of the region combined, fears were high that if the Golden State were left to its own devices, it would dominate and control the Colorado River Basin in perpetuity. In response, a lawyer from the state of Colorado named Delph Carpenter authored a bill, passed in Congress and the legislatures of all seven states within the basin in 1921, that created a commission of delegates tasked with forging an equitable agreement over the course of the next year for how to plug the mighty river so its water could be put to use in extending the nation's imperial grip across the magnificent landscape it had claimed.

Ahead of an initial series of meetings in Washington, D.C., the governors of Arizona, California, Nevada, Utah, New Mexico, Wyoming, and the state of Colorado each appointed an official representative for their claim to the river. After the first few sessions ended in deadlock, the commissioners took a seven-month break before reconvening in November at Bishop's Lodge, a resort in the Sangre de Cristo Mountains north of Santa

Fe that had been built on the site where the French missionary Jean-Baptiste Lamy first settled in the 1850s.

To kick off the negotiations in Santa Fe, the commission's chairman, Herbert Hoover, called on representatives from each state in alphabetical order, asking for their vision for how the Colorado's water should be divided. Arizona's Winfield Norviel went first, suggesting a system based on current use, with shortages shared equally across the basin. W. F. McClure, from California, asked to defer his turn, ceding the table to Delph Carpenter, who was representing Colorado. Carpenter proposed breaking the river into two basins. The senior rights of irrigation districts in Southern California would be respected, even as the thinly populated Upper Basin states of Colorado, Utah, New Mexico, and Wyoming would still be guaranteed a share of the river.

R. E. Caldwell, from Utah, offered some tweaks to Carpenter's plan, namely that the dividing point be moved north from Yuma, Arizona, to Lee's Ferry, the site of a historical ford across the river near the Utah border. This would put Nevada in the Lower Basin, along with California and Arizona. Then California's McClure offered the floor to George Hoodenpyl, a Long Beach lawyer whom he'd brought along as an advisor. It was Hoodenpyl who suggested that, rather than work out a complex water allocation scheme on the spot, each of the two basins be given rights to half of the Colorado River's water, which could then be divided internally.

Representatives from the thirty tribes within the basin were not invited to the negotiations in Santa Fe. It took until day twenty of the deliberations for the question of tribal water rights to even come up, with Hoover proposing adding a "wild Indian article" to the treaty that read, "Nothing in this compact shall be construed as effecting the rights of Indian Tribes." His suggestion was immediately shot down. With no Utes from the Colorado's headwaters, O'odham from its tributaries, or Hualapai from its Grand Canyon present to protest, the state delegates adjourned for the evening and did not return to the topic until it was time to finalize the treaty.

With all the delegates having come to a general agreement on a framework for the compact, a representative from the Bureau of Reclamation estimated that the Colorado River's annual flow averaged out at 16.5 million acre-feet, with each acre-foot equivalent to about 326,000 gallons, the quantity of water needed to flood an acre of land to a depth of one foot. The estimate of 16.5 million acre-feet, however, completely disregarded the brutal droughts of the late nineteenth century that had crippled agricultural production in Arizona and Southern California—a critical accounting error that, a century later, the basin is still struggling to correct.

For the next two weeks in Santa Fe, the delegates went back and forth on the details, before finally agreeing on November 24 to a deal whereby each basin was guaranteed 7.5 million acre-feet of water, with the Lower Basin being entitled to grow its use by an additional million acre-feet per year in the future. As far as Mexico was concerned, "as a matter of international comity" that country would be assigned whatever excess water was left over if and when the United States deigned to share the Colorado with the people who lived in the river's delta. Native Americans were completely left out of the final agreement, which reserved only one line for them: "Nothing in this compact shall be construed as affecting the obligations of the United States of America to Indian tribes." A wishy-washy revision to Hoover's original suggestion, the new language eliminated the suggestion that tribes had any rights at all.

Announcing the deal, Hoover proclaimed, "The foundation has been laid for a great American conquest. The harnessing of the giant Colorado River will follow the ratification of the pact by the seven states of the Colorado River Basin. With such ratification, the next step will be the construction, without delay, of a control dam, under authorization of Congress. Then the Southwest will come into its magnificent heritage of power and life-giving water, and all the nation will be vastly benefited."

The federal embrace of reclamation proved a godsend to the farmers of the Salt River Valley. But while reclamation stabilized agriculture in the region, the initial set of projects like the Roosevelt Dam provided little aid to Phoenix's burgeoning residential population, which was left to pump groundwater, much of it silty and undrinkable. By 1922, the city had grown enough that wells alone would not suffice, so it looked northeast, thirty miles past the McDowell Mountains to the supposedly underutilized Verde River. A pipeline of redwood was built between the river and the city's water mains on 12th Street with the capacity to convey 15 million gallons of water a day. Within a decade, a new, concrete pipeline was necessitated, this one large enough to carry 30 million gallons.

A new drought set in shortly after the second's pipeline's completion, and it grew so severe that, by 1940, Lake Roosevelt had completely dried up, rendering the dam nothing more than a wedge of concrete between a dry gulch and a bed of silt. Lin Orme, the longtime head of the Salt River Valley Water Users' Association, who had come of age during the region's wettest years and been just nineteen during the catastrophic flood of 1891, told everyone who would listen that the rain would come: "All you have to do is have faith." As promised, the rain arrived that winter, leading some fifty thousand revelers from throughout the Valley to parade through downtown Phoenix toward a stage erected in front of the Heard Building, which showcased a twenty-four-foot-tall diorama of the Roosevelt Dam, complete with water gushing through the model's spillways.

The relief was short-lived. The dry times returned, yet Phoenix kept growing, leading to the inevitable point at which the Verde no longer proved sufficient to meet the needs of the city. Reservoirs were built and new wells drilled, but nevertheless, by 1946, Phoenix's use was outpacing its supply by a million gallons of water per day. That July, municipal officials secured enough water to meet the shortfall from the Roosevelt Irrigation District. It wasn't enough. Within five years, the city was diverting more than twice the amount of Verde River water it was entitled to. The situation came to a head on July 3, 1951, when water officials informed the city that only a ten-day supply remained in the river.

The Arizona Republic's alarming headline the next morning— Independence Day—read "Phoenix Water Crisis Near." In the following days, the paper's columnists sought to quell the panic by reassuring the city that the shortage in the Verde would easily be met by pumping groundwater, never mind its depth or palatability. Anyone who appealed to Orme for guidance heard the same thing he'd been saying for years: "It will rain—it always has." Though the entire history of Phoenix up to then had been one of scarcity, the notion that it would definitely run out of water was impossible for planners to entertain with any degree of seriousness.

It was at that point—the most acute moment of crisis the young city had yet experienced—that the miraculous transpired: Orme was right. Again. Five inches of rain fell in August, more than half of the precipitation Phoenix receives in an average year. It's a monthly record that still stands. The reservoirs were refilled, the groundwater recharged, and Phoenix had bought enough time to broker a deal with the surrounding farmers to deliver water from the Roosevelt reservoir to former agricultural land within the city limits that had been redeveloped into subdivisions and office buildings. As the city continued to grow, more and more farmland was swallowed up and the Salt River gradually shifted from supplying an Edenic bounty of cotton, oranges, and dates through lush irrigation channels to maintaining an underground spiderweb of plumbing lines that serviced strip malls and spas.

Raymond Carlson was born in the mining community of Leadville, Colorado, in 1906, though his family moved to Arizona when he was still a child. They, like so many others, were lungers: Carlson's father had contracted a respiratory disease, so the family sought the "climate cure" of the desert. They ended up in the small town of Miami, eighty miles east of Phoenix, where Carlson's father found work mining copper instead of silver.

Carlson excelled in school, earning admittance to the University of Arizona and then transferring to Stanford, where he wrote for the *Chapparal*, a humor magazine, and graduated Phi Beta Kappa. He had hoped to go to law school but couldn't afford the tuition, so he returned home to Miami and committed himself to writing instead. The local newspaper, the *Miami Silver Belt*, offered him a job, but that only lasted a few years—the paper shuttered in 1934, another casualty of the Great Depression.

Afterward, Carlson survived as a waiter and bookkeeper, before finally getting his big break: his wife's uncle, Rawghlie C. Stanford, was hunting for a speechwriter. Stanford had been defeated in the Democratic primary during his first run for governor but prevailed in 1936—and just like that, the ambitious young journalist had a direct line to the most powerful office in the state.

Early on, the masthead of *Arizona Highways* had been remarkably unstable. Each new governor installed his own editor of the periodical, and sometimes more—Stanford's predecessor employed three different editors across his four years. Not that the magazine that emerged from all the political jockeying was of much interest to anyone aside from civil engineers. The 1921 legislation that apportioned funds for Arizona's department of roads to operate a magazine imagined a publication that would "encourage travel to and through the state of Arizona," and the bureaucrats who initially oversaw *Highways* took the name literally, printing maps of the state's thruways and articles explaining the numbering schema for its roads.

After Stanford named Carlson editor in 1938, he resolved to reinvent *Highways* as a tourism magazine, one that did not just explain how to travel across Arizona but also would persuade readers that they should go there in the first place. At the same time, he proved much more politically adept than the previous heads of *Highways*, putting distance between the magazine and the Highway Department while still staying in the good graces of elected officials. Even after Stanford declined to stand for reelection in 1939, Carlson would remain in control of *Arizona Highways* for another three decades, his editorship interrupted only by his service in World War II.

Part of the task Carlson set himself was arguing that the deserts of Arizona weren't nearly so desolate as his readers believed. In the early 1940s, he penned an essay titled "Water in the Magic Land" that was accompanied by photographs of Arizona's many early triumphs of reclamation, including the Bartlett Dam on the Verde, which Carlson called a "poem in concrete." "The judicious use of water, its conservation and the use of the soil in relation to the water that comes to it—that will be the yardstick generations of scholars of other civilizations will use to measure the greatness or the littleness of our civilization," Carlson wrote, before turning to the place he now called home. "Out of what was once a forlorn desert, arises the great city of Phoenix, Arizona—much of whose greatness, much of whose rich promise is the result of water in the magic land. The decades to come will add to the stature and importance of this city as they will add to the richness and greatness of this, our Empire in the West."

During his time in the Army, Carlson kept mainly to himself, seemingly too preoccupied with his plans for *Highways* to fully engage with his fellow soldiers in Southeast Asia. Which isn't to say he came off as shy. One remembered his habit of stripping nude to lounge in the sun like a lizard, as if its rays alone could transport him back to the Sonoran Desert. In 1945, Carlson returned home from his deployments in Luzon and New Guinea with a renewed sense of purpose. In his first editor's note since V-J Day, Carlson wrote that "having been away from home, one finds Arizona more attractive in every way—the skies bluer, the desert dreamier, the distant hills more purple, the lure and call of the lonely places more real and poignant." Rather than merely appreciating the spectacle, Carlson felt compelled to spread the good word. "Taking up again the job of telling of Arizona in these pages is now more than a job," Carlson resolved. "It is a crusade."

Carlson was soon inducted into a fraternity that was already hard at work turning Phoenix into a locus of American life in the West to rival Los Angeles and San Francisco. Most prominent among this circle was Walter Bimson, the banker whom *Highways* called "Arizona's Indispensable Man." Bimson's Valley National Bank was one of the few in the state to survive the Great Depression; by the 1940s, it controlled roughly 70 percent of deposits in Arizona. Among Bimson's closest associates was the real estate developer and future New York Yankees owner Del Webb, as well as Barry Goldwater's brother Robert, a major force in the local chamber of commerce. Then there was Dwight Heard, a Chicagoan executive at one of the nation's largest tool companies who had first moved to Phoenix in 1895 after a doctor recommended the climate as a treatment for his respiratory disease. After he recovered, Heard bought real estate, amassing more than 7,500 acres in his first few years as an Arizonan. Over the ensuing decades, he established a cattle company, a newspaper, and a real estate brokerage.

The efforts of these men to turn Phoenix into a boomtown and then reap the profits paralleled the activities of the booster networks that championed the meteoric growth of other Sunbelt cities like Tampa, Dallas, and San Diego. What makes their success so miraculous is that Arizona, up until World War II, was understood by most Americans to be a uniquely hostile place, making Phoenix a far harder sell than a city in the South or on the Pacific Coast. That was where Raymond Carlson, with his keen sense of how Arizona's romantic landscape could be leveraged to support its growth, came in. Back from his military interregnum, he wrote of how "Mr. and Mrs. America, this summer free as the wind and completely unfettered by restrictions and worry of war," might find themselves drawn to Arizona, where "the horizons dance in the sunlight, alluring hoydens, full of promise and things to come." The vision of the state he sought to package and distribute around the country was not the rugged, wild desert of John Ford films or Zane Grey novels. It was a more alluring place, a land of respite and wonder.

At the same time, Carlson had a powerful sense for how the tastes of readers were shifting with the times. As popular culture progressed into visual media, with radio giving way to film and then television, photography became the magazine's calling card. *Highways* was the first consumer publication to print an all-color special edition, in 1946. Other issues from that era included a dazzling image of Navajo National Monument, an Ansel Adams photo of Monument Valley, and a vertiginous shot of a waterfall pouring through Havasu Canyon—all of them made, in Carlson's words, "more real and poignant" by the addition of color.

The magazine's circulation at the time of Carlson's arrival was around ten thousand; by the 1950s it reached two hundred thousand. Readers couldn't get enough of the full-color landscape photography that Carlson published with increasing frequency, and which set *Highways* apart even from the likes of *National Geographic*, which did not begin putting photography on its cover until 1969. Carlson worked with many of the period's most renowned nature photographers: Esther Henderson shot a number of features on the Petrified Forest and gold-leafed paloverde trees; David Muench captured the brilliant, fleeting bloom of an ocotillo; and Norman G. Wallace framed the Superstition Mountains behind a stand of saguaros.

Though Arizona's neighboring states had plenty of their own dreamy sights and astounding natural landmarks, New Mexico, Nevada, and Utah lacked official media organs capable of competing with *Highways*. Consequently, tourism in Arizona soared as it increasingly became synonymous with the entire Southwest. In the three decades after World War II, annual spending by visitors to the Valley of the Sun spiked from $30 million to more than $1 billion. Some of the guests were so taken with Arizona they never left.

The conversion of tourists into residents didn't happen by accident. One pamphlet issued by the Phoenix chamber of commerce described the many attractions of the state, before concluding, "Make the Valley of the Sun your home base for discovering even more about Arizona." Del Webb's Sun City, the development that pioneered the idea of a community exclusively populated by retirees, issued a "vacation special offer" to prospective residents, where they could pay seventy-five dollars for a week's stay that would include the opportunity to "meet the wonderful, friendly people who live there" and "tour the furnished model homes and apartments available for sale." By 1960, the state's population had tripled from the prewar years, to 1.3 million.

Inez Robb, a nationally syndicated columnist, called *Highways* "habit forming." If the magazine exuded "a faint tincture of snake oil" to skeptics in the Northeast, Robb wrote, it was only because "Arizona shares with the giraffe the distinction of having to be seen to be believed." It wasn't just Arizona's mind-bending landscape and flora that commanded the attention of readers like Robb; it was the astounding feats of human ingenuity that were enabling the state's population boom. As Carlson wrote, "The handiwork of the engineer is felt in every phase of our environment that makes living genteel, safe, and comfortable."

In issues from 1957 and 1964, the magazine traded its usual cover shot of a natural landmark for downtown Phoenix, with the latter edition describing the state's capital as "the miracle city in the sun." Inside were aerial photos of a neighborhood shaded by palm trees and a newly paved freeway with no more than a handful of cars on it, extending into the mountain-studded horizon. One article that dubbed the city the "economic capital of the great Southwest sun county" bragged about Arizona State University's three-hundred-acre campus in Tempe and the eighty-seven shopping centers in the metropolitan area; another lauded its parks, art museums, and ballet school. In his editor's note, Carlson mused, "What the future holds for the community staggers the imagination not only of the statistician but of the dreamer and poet." Subscribers responded to these exhortations by descending on the city. As Robb wrote, once you got "hooked" on *Highways*, "You begin to believe and then you want to go, go, go."

All of this complemented the efforts of Phoenix's booster class to persuade national corporations to expand their operations to the city, as exemplified by the special events committee of the local chamber of commerce. The group was founded in 1937, three years after a local advertising agency coined the moniker "The Valley of the Sun" as an inviting alternative to the hardbitten Salt River Valley. The event committee's main charge was squiring visiting executives around town, though it also planned special tourist draws, like the annual Fiesta del Sol. For a name, the group chose Thunderbirds because they got a kick out of play-acting Indian, with costumes of turquoise-studded belts, velvet tunics, and heavy, silver medallions. They even called each other names like Big Chief, Sachem, and Medicine Man—titles that persist to this day.

Bob Goldwater was one of the earliest members of the troop and the first organizer of what became its signature event: the Phoenix Open golf tournament. Bob and Barry's mother was fond of golfing, and the two boys were regulars at the Phoenix Country Club growing up (the Goldwater family had helped found it). When the chamber of commerce was casting around for ideas that might lure more visitors in the late 1930s, Bob and the other Thunderbirds suggested a golf tournament scheduled for midwinter, in hopes that, as the *Phoenix New Times* put it, "the name Phoenix would become synonymous with winter salvation through golf."

"It was all timing," Bob told the *New Times* in 2001, when he was ninety years old, "My mom grew up in western Nebraska and went to school in Chicago and she would talk about how tough this time of winter could be on your brain. It's a rough time to hear about people playing outside in the sun. We just figured people couldn't help but come be a part of it." The Phoenix Open eventually drew stars like Bing Crosby and Bob Hope, whom Goldwater entertained with the rowdy parties he threw at his mansion on the course's third fairway. One of the Thunderbirds' responsibilities at the tournament was playing pranks, like stashing a speaker in the cup on the eighteenth green to throw off contestants. The message was clear: Arizona was *fun*. Or, at least, fun for the sort of person who relishes eating hot dogs at Augusta.

The efforts of the Thunderbirds complemented the corporate-friendly policies of Arizona's political leaders, who restrained union organizing and repealed a sales tax on products manufactured locally for sale to the federal government—a major boon to defense contractors. When, in 1955, Sperry Rand was scouting locations for a new aviation electronics plant, local businesspeople raised $650,000 to buy a factory site for the company ahead of a planned visit from its executives, which the Thunderbirds offered free of charge over cocktails at the Phoenix Country Club. Sperry Rand bit, as did Motorola, Lockheed Martin, and Honeywell, all of them setting up major facilities in the Valley of the Sun. Between 1940 and 1963, the manufacturing output of Phoenix ballooned from under $5 million to more than $435 million.

Despite his important role in fostering Phoenix's expansion, Raymond Carlson was always too much of a romantic to fit in with the money-mad Thunderbirds. Even as men like Bimson, Heard, and Webb made millions off the growth of Phoenix and politicians like Barry Goldwater and Carl Hayden became Washington stalwarts, Carlson remained ensconced at the *Highways* office, a true believer that "telling of Arizona" was a righteous cause.

For all its national impact, until the late 1960s the magazine's editorial staff consisted entirely of Carlson and George Avey, its longtime art director. Robert Stieve, when we spoke, seemed as enthralled with Carlson's editorial vision as he was blown away by his predecessor's ability to churn out the magazine every month with so little assistance. "I look back at these old issues, and it was just two guys doing it. Ray and George, sitting down, finding these stories, pulling them together." That dedication to the magazine was admirable, though unsustainable. In the 1971 issue announcing Carlson's retirement, his successor, Joe Stacey, shared the news that the editor had been sick for more than two years, "during which time it was necessary for Mr. Carlson to direct the editorial details of the magazine from his home and at times from his hospital bed."

Carlson's wife had died a few years before he left *Highways*, so, as his health deteriorated further and he began using a wheelchair, he moved from his Wright-designed bungalow to a nursing home in Scottsdale. Contrary to his embrace of the leisurely life in his writing, Carlson wore a tie every day of his retirement and occupied his time with reading and crossword puzzles rather than the sun-worshipping of his youth. His profits from the society he helped to dream into existence were few; because he and his wife had no children, when Carlson died in 1983 his entire legacy was *Highways* itself, a magazine that, over the coming years, would become as singularly preoccupied with its old editor as it was with Arizona's landscape. The editorial staff's enduring awe at Carlson's triumph would be refracted again and again in its regular anniversary issues and revisitations of his career until the magazine became a kind of echo chamber, *Arizona Highways* celebrated as regularly and vociferously as Arizona itself.

After my conversation with Robert Stieve, I drove due east on Indian School Road through Encanto Village until the avenue began to swoop in parallel with the Arizona Canal, which was built in the 1880s and is one of the handful that do not follow the ancestral Sonoran blueprint. After twelve miles, I reached Old Town Scottsdale, a purportedly historic shopping district where luxury stores and salons were organized along squeaky-clean, colonnaded arcades. It was a bluebird February day, the sky clear and just a hint of heat in the air. After ordering a cappuccino from a barista behind an elegantly tiled bar, I crossed back over the canal on a pedestrian bridge to an apartment complex called the Broadstone.

While for most of its length the Arizona Canal's banks are lanes of sunbaked earth wide enough to facilitate access for the Salt River Project's utility trucks, here the waterway was lined with palm trees and eucalyptus that framed the view to the north of the coarse granite slopes of the McDowell Mountains. In front of the apartment building, the Scottsdale Public Art Program had installed a sculpture that captured the "western spirit of adventure" by depicting two Pony Express riders on galloping steeds. Contrary to its appearance, the sculpture was not meant to commemorate the Pony Express itself, but rather the "Hashknife Pony Express," an annual tradition begun in 1958 in which men dress up in nineteenth-century garb and embark on a two-hundred-mile ride from Holbrook to Scottsdale. When they arrive, the riders are greeted with the "Parada del Sol Parade and Trail's End Festival," one of the tourism board's signature events.

Unmentioned on the plaque was the important detail that the actual Pony Express never passed through Arizona. The route was strung between Missouri and San Francisco. Unlike the old Entrada pageant in Santa Fe, in which a sanitized version of Diego de Vargas's reconquest of New Mexico was canonized, there was no actual history of the Pony Express in Arizona to misconstrue. Nevertheless, the Hashknife Pony Express is among the state's most storied traditions. In 1988, *Arizona Highways* wrote a glowing feature on what it called "the legendary Hashknife Pony Express." Two years earlier, the governor designated "Pony Express Week" on the official state calendar. A member of the Arizona Historical Society crowed to *Highways*, "These Hashknife men have truly succeeded in capturing the flavor and spirit of the pioneer mail riders." All of this left the sculpture with something of a mise en abyme effect: it was a monument to the ability of the Valley of the Sun to generate tourist interest out of nothing at all.

For all the astounding things Phoenix has achieved in the 150 years since its founding, the city's culture has remained rooted in an earnest and unwavering belief in its ability to forge a vision of paradise out of whole cloth. That this vision mostly appealed to a particular sort of person—wealthy, white, and nearing retirement—was the whole point, and it was why the city had been able to surpass the settlements of the old Southwest. From the very start, Phoenix offered itself to the rest of the nation as an empty page, a place with no history where new arrivals could compose their own story.

For many Anglos, I realized, Albuquerque's Plaza Vieja might be worth a visit, but its adobe buildings and the high likelihood of overhearing someone speaking in Spanish—let alone Diné or Tewa—meant the place was too dissimilar from their origin point to ever become home. Old Town Scottsdale provided an alternative Southwest for those who found just being in the desert jarring enough. The open-air mall, as bland and luxurious as what can be found anywhere in the Sunbelt, carried the implicit promise that shoppers would never be forced to confront the fact that there was ever a world before American empire.

Rather than follow the straight shot of I-10 across the dry plain that separates Phoenix from Tucson, I took the back way: Route 60 toward Apache Junction. On the outskirts of town, I passed a soon-to-be housing development along Ironwood Drive. Earth movers and backhoes circled a flat clearing with a grove of transplanted juvenile trees, neatly lined up in their planter boxes as if waiting for their turn to be deposited along a cul-de-sac of the future.

I drove southeast, the Superstition Mountains to my left. Somewhere deep in that jagged, pyroclastic massif was the Lost Dutchman Mine. Early in the nineteenth century, an Apache band camping around the Verde River is said to have escorted a white doctor who cared for them to a secret lode in the mountains, where he was invited to take as much gold as he pleased. Though it was a frigid winter in the high desert, the doctor stripped off his long underwear, knotting the legs and waistband into a bag he used to carry out a fortune in golden nuggets. In the 1870s, a German prospector named Jacob Waltz was hunting for the same mine when he came upon three Mexicans around a campfire. After Waltz talked his way inside the little cleave on the hillside they were guarding, he saw a glittering vein a foot and a half wide. He massacred the trio with a shotgun and claimed the gold for himself. Waltz would kill four more men to protect his windfall until, lying on his deathbed, he took a younger man named Dick Holmes into his confidence, doing his best to explain how to find the mine that he had become too infirm to visit himself. The only proof he could offer Holmes was a tin box under the bed that was filled with forty-eight pounds of gold. Dick Holmes spent his life haplessly searching for the mine. His son, Brownie, did the same. To this day, the gold mine's location remains a mystery.

I peeled off onto State Road 79, following it south over the Gila River and through Florence. The first indication that I'd reached the town was the razor wire topping the fence around the medium-security prison. More than 3,500 people were incarcerated at the seventy-three-acre complex, many of them migrants, since one of the privately owned Correctional Center's primary clients is Immigration and Customs Enforcement, better known as ICE. The massive facility had made Florence a beacon in the constellation of towns across the Southwest whose economies have come to revolve around the militarization of the border, with the region's federal workforce seeming to grow by the year.

To my astonishment, the further south I got the greener the landscape became. It was the tail end of an unusually plentiful monsoon season, and both sides of the road were lined with lush grass, tall enough to nearly obscure pervasive cholla cactus, all of which looked almost brown in comparison. In the distance, the blacktop unspooled directly into the nine-thousand-foot summit of Mount Lemmon, its top obscured by a dense bank of clouds. The full expanse of the Santa Catalina Mountains clarified after I passed Oracle Junction, their rocky slopes wrinkling away from Mount Lemmon as blue sky filled the shoulders between each peak. Then, the banks of paloverde trees on either side of the road began to be interrupted by palms and the dreamily named subdivisions of far northwest Tucson. Bonanza Heights. Rancho Vistoso. Garden of Eden.

RECALCITRANCE

Only the sunlight holds things together. Noon is the crucial hour: the desert reveals itself nakedly and cruelly, with no meaning but its own existence.

—EDWARD ABBEY

Ofelia Zepeda occupied an old, high-ceilinged office on the ground level of a century-old brick building at the University of Arizona. She wore a necklace of beaded coral, and her hair rolled gently over her shoulders, willowy and white. Behind her desk, she'd decorated with a print by the Hopi artist Dan Namingha and a map of Arizona, detailing its waterways, both of which were lit softly by the glow of clerestory windows.

Zepeda is Tohono O'odham but grew up away from the reservation, in a cotton-farming community near the remains of a settlement known as Casa Grande, which sits about halfway between Tucson and Phoenix and is thought to have been built in the fourteenth century. This means she spent her childhood in close proximity to the historic trade route that connected the farms of the Salt, Gila, and Santa Cruz Rivers to fishing camps along the Pacific Coast, and from there to Nahua-speaking peoples in the jungles of Central America. Evidence of these continent-spanning ties persist in archeological sites across the Sonoran Desert in the form of iridescent shells, macaw feathers, copper bells, and ornate, reflective discs decorated with a mosaic of iron pyrite. Despite being separated from the Aztecs by more than a thousand miles, the ancestral peoples of the Sonoran Desert even played a version of the game ullamaliztli, where a heavy rubber ball was knocked around an earthen oval court using hips, knees, and elbows. More than two hundred of these ball courts have been found across southern Arizona.

Much of that heritage was invisible when Zepeda was growing up in the 1950s. Her family was surrounded on all sides by the cotton fields that then dominated agriculture in Arizona, ultimately smothering four hundred thousand acres of the state. Almost all of the crop was hybridized Egyptian cotton that had been introduced to Arizona around the turn of the twentieth century rather than the Sacaton variety that Zepeda's ancestors had been cultivating for millennia. "Most everybody worked as farm laborers," she remembered, "that's how you had to provide for yourself and your family." In the summertime, cotton fields would gradually turn the orange earth of the vast expanse around Casa Grande green, before exploding into white every fall. The only sound out in the fields was wind rustling cotton bolls and the occasional whirr of a passing truck.

"We were one of those families where no one had any experience with any kind of schooling," Zepeda said. When she first enrolled at a rural public school, she didn't speak any English. Nevertheless, she excelled. Her brother was equally bright, particularly at math and art. "He was one of those people that had that natural gift," she said, but since their parents needed him to work, he dropped out after only a year or so of high school. "I have other cousins that are like that, but they never went to school." Zepeda described these circumstances in a measured voice. There was no obvious regret, just a recognition of the way things went for so many O'odham kids, then and now, no matter their aptitude for education.

In the end, Zepeda became the first member of her family to receive a high school diploma. This she attributes to chance. "I just happened to be in a school where a teacher found me, or I found a teacher. In that era, there were federal programs to increase the number of minority students going to college. So myself and my cousin, a counselor put us in one of those pro-

grams and we got pre-college training. We got on that track. My
cousin, she went to a technical school, and I went to university."

Arriving in Tucson was a shock compared to the rural environment she was used to: beyond the sheer size of the city, which
seemed to be growing by the day, subdivisions spreading rapidly
away from the university and toward the Santa Catalina Mountains, there was the neon madness of Speedway Boulevard. Soon
after Zepeda started college, *Life* magazine called Speedway
"America's ugliest street" because of the visual cacophony created
by its enormous, overbearing signs: Saxon's Sandwich Shoppes
("Our Food is Rated G: Great"), Precision Motors, Drive-In
Liquor Window, Andy's, Golden Pheasant Restaurant, Ranch
Center.

Soft-spoken and patient, Zepeda buried herself in books. In
quick succession, she earned a bachelor's, master's, and doctorate
degree. Though originally interested in sociology, Zepeda was
convinced to try linguistics by Kenneth Hale, a scholar at MIT
who had grown up in Tucson and been invited back to serve as
a guest professor for a few semesters. To Zepeda's surprise, the
charismatic Anglo who wore a prominent belt buckle commemorating his victory in a high school bull-riding competition spoke
excellent Tohono O'odham. He explained to her that he'd picked
up the language from a teenage friend (a famous polyglot, Hale
was said to be able to communicate in dozens of tongues, including Diné, Hopi, Wampanoag, and the Australian Aboriginal language of Warlpiri). Recognizing that few academic surveys of
the Tohono O'odham language had ever been undertaken, Hale
encouraged Zepeda to study the formal components of the language she'd been speaking for her entire life.

"I didn't realize what I was doing was special," Zepeda told
me. She described the generations she has instructed in O'odham
since publishing *A Tohono O'odham Grammar*, both university
students and reservation teachers who hoped to develop literacy
in and then share their mother tongue. "I saw it like a job. I didn't
think what I was doing was unique until much, much later when
I'd meet people I worked with early on—the impact of being able
to read and write their own language."

With only around fifteen thousand speakers in the United

States and Mexico, O'odham is designated as an endangered language. Zepeda's work has helped create a framework for its restoration. The Tohono O'odham Legislative Council began building a language center at a tribal community college in 2020 and the language is also offered at Ha:ṣañ Preparatory & Leadership School, a charter high school in Tucson's Rincon Heights that serves Indigenous students from across the city, along with commuters from the Tohono O'odham reservation (whose largest town, Sells, is about an hour away).

While Zepeda's manner had initially felt to me as somehow out of sync with the hustle of the university, the longer we spoke, the easier it was to understand her as an integral figure in Tucson. Her whole demeanor suggested immovability, centrality. "*Tohono* just means 'desert,' 'desert people,'" Zepeda said. "That's who you are, what you were made to be. You can't get away from it." Outside of academia, she has published three poetry collections, much of her creative work focusing on what it means to be not just in the Sonoran Desert, but of it. "Cuk Ṣon is a story. / Tucson is a linguistic alternative," she wrote, in a poem called "Proclamation":

> *The true story of this place*
> *recalls people walking*
> *deserts all their lives and*
> *continuing today, if only*
> *in their dreams.*

Was it possible, I wondered, for someone outside the lineage that dates back to the ancestral Sonoran Desert peoples to really belong here, in a landscape characterized by such profound extremity? Could those who have sought to master the Sonoran Desert with air conditioning and aqueducts really call it home?

Zepeda responded to that line of questioning with typical equanimity. She described an annual ceremony the Tohono O'odham perform when the monsoon rains of late summer hit, which seeks to cleanse the earth and "prepare it for another cycle." Before European contact, she said, this ceremony "was for the O'odham and everything around us, the animals and all the other people that might be somewhere on the earth. After contact, all those people are included. You cleanse the earth for everybody. When things change dramatically—it doesn't rain, the patterns change—they recognize it and say, 'We're not living right.'" That sense of continuity between the personal and global, she said, has only become more salient in recent years. "With the droughts and wildfires, people will say, 'We're not living right.' Not just the O'odham, but everyone."

After passing through the 6th Avenue underpass into downtown Tucson, the first building on my right was covered with murals. There was a "Black Lives Matter" banner and a portrait of Che Guevara, along with a bright yellow wall depicting a trio of women with sugar-skull masks under the legend "mexican@s," with "Las Adelitas" and "¡Viva la mujer!" inscribed below. In the middle of the array, "You are on O'odham Land" was written in varsity font, with a portrait of an old woman with a walking stick accompanied by white lettering that read, "We are still here!"

The necessity of this assertion of belonging became obvious the deeper I walked into downtown. Tucson's origin as an outpost of the Spanish Empire is particularly legible along Meyer Avenue and Court Street in El Presidio, a designation dating back to the founding of a fort there in 1775, less than ten miles up the Santa Cruz River from the San Xavier del Bac Mission, established in 1700. The Spanish needed a defensible position, as Apache raiding parties had moved to the region after they were pushed out of what's now New Mexico and Texas by the Comanche. In short order, the Presidio accumulated a population of more than a thousand displaced Spanish and Tohono O'odham farmers, but it fell into decline after Mexico won its independence in 1821. The new government was simply too overstretched and disorganized to offer much protection to a small settlement some 240 miles from Hermosillo, the capital of Sonora.

As I crossed Alameda, the sense of Tucson's Spanish and Mexican past evaporated into the blinding light, as charming old adobe bungalows were replaced by modernist high-rises that accommodate a full suite of city, county, state, and federal agencies, from the Tucson Water Department to the U.S. Marshals. The streets of El Presidio were well planted with shade trees; by contrast, the wide plazas between the office buildings baked in the sun.

I felt the disjunction in the urban fabric like an electric shock. It sent me crisscrossing the brickwork square, searching for some explanation, but the only historic marker I could find was a monument to the Mormon Battalion outside the city hall. This was the group of five hundred LDS faithful who volunteered to fight in the Mexican-American War and briefly stopped in Tucson in 1846, midway through their two-thousand-mile march from Iowa to San Diego. After the commander of the local Mexican forces retreated to the mission of San Xavier del Bac, the Mormon Battalion was greeted by around a hundred residents, for whom the war was incidental—their livelihood depended more on the climatic swings that determined the productivity of their farms than on the nation that technically controlled the Presidio. Squinting to read the bronze plaque in the brilliant sunlight, I learned that the Mormon Battalion had been the first to claim the city for the United States, though it would ultimately revert to Mexico until 1853, when the United States purchased the land south of the Gila River from its defeated and exhausted neighbor, solidifying the southern border that persists to this day.

From the Gadsden Purchase until the 1960s, the urban core of Tucson was synonymous with Arizona's Mexican American community. As early as 1867, a newspaperman in Prescott warned Anglos against migrating there, writing that "Tucson is no place for a poor white man, as all the work is done by runaway peons from Sonora, who arrive there by the hundred in destitute condition, and who will almost work for board." The Anglos who did arrive congregated near Amory Park; after the Union Pacific Railroad reached Tucson in 1880, white neighborhoods began to spread north and east of the tracks, where the campus of the newly founded University of Arizona was likewise designated in 1885. Meanwhile, the old barrios grew to the south and west, establishing a pattern of segregation that is still in place.

As Tucson expanded, the area south of El Presidio—its various barrios collectively known as La Calle—remained a draw for Mexican Americans. People came from barrios all over town to see Spanish-language films at the La Plaza Theatre on Congress Street, a venue that also hosted the singer Luis Pérez Meza and the Argentine dancer Libertad Lamarque. Down the street, the original office of the Alianza Hispano-America hosted gatherings on the first floor that were known to get so wild that some residents called the space the Spanish word for "jungle," La Selva. While public accommodations in north Tucson were strictly segregated, La Calle was a place where Black residents could sit wherever they wanted at the movie theater or drink in peace at a soda fountain. In addition to all the businesses owned by Mexican Americans, there were a number of Chinese grocery stores, a clothing shop operated by a Syrian family, and Jimmy's Chicken Shack, which occasionally held events for Black entertainers from Chicago and Los Angeles.

However vibrant La Calle was, the growing Anglo population of Tucson viewed a nonwhite neighborhood in such close proximity to downtown as a liability. In 1931, a reporter from the *Tucson Daily Citizen* wrote that visiting the area was akin to "crossing the international line at Nogales . . . As we proceed down this bizarre old world *rialto* we realize that we have left the land of the hot dog for the land of the *chile con carne*, the land of the go-getter for the land of *mañana*." Two decades later, the architect Edward Nelson groused, "The industrialist who visits Tucson will have a bad impression of the city if he sees slums such as these so close to the downtown businesses." His choice to call La Calle a "slum" had less to do with the conditions of the buildings than with their occupants, as, according to an analysis the city commissioned in the 1940s, one of the two criteria for declaring an area blighted was simply that it featured an "intermixture of racial and ethnic groups." The other criterion was decrepit housing.

Though most homes in La Calle were in at least serviceable condition through the 1950s, city officials quietly fostered the neighborhood's decline for more than a decade, heeding the chamber of commerce's call in 1939 to embark on a "federally-assisted slum clearance program." Only 20 percent of homes in the area were owner occupied and code violations of rental units went largely uninvestigated by the city, no matter how much their tenants complained.

In 1958, the city published a plan to redevelop 392 acres of downtown Tucson, including practically all of La Calle. In the years that followed, building owners in the district were told that making improvements to their properties "would be a step against the tide in the area." When one woman who owned a few apartments sought a permit to upgrade their plumbing, she was denied, while other homeowners struggled to get loans to improve their properties, given that the entire neighborhood had been redlined. In 1964, the city stopped collecting trash in La Calle, ignoring calls from residents asking for their previous service to be restored. As an urban renewal official named Don Laidlaw later recalled to the *Tucson Weekly*, starting in the 1950s, "there was some zeal on the part of building inspectors to rein in improvements, because the conclusion had already been reached that everything was going to get torn down—why spend public money on buying back improvements?"

In March 1966, the city held a bond issue referendum to raise the $14 million needed for the Pueblo Center Redevelopment Project, which promised to level the eighty-acre core of La Calle that a city councillor named Kirk Storch described as a "cancer . . . spreading over the entire downtown." At the time, Arizona law stated that only property owners could vote in municipal bond elections, so only a fraction of the neighborhood's residents had any say in its fate. Less than 4 percent of the city's population voted on the measure, which passed by a mere three thousand votes. Demolition began a year later; by 1969, more than two hundred buildings had been destroyed, displacing at least seven hundred residents and a hundred businesses. The original 1958 proposal had been derailed by its inclusion of public housing, so none was included in this plan, an oversight justified by one of its proponents on the grounds that Tucson already had "adequate low-cost housing" elsewhere. Come 1971, the former residents of La Calle had been scattered across the south side of town and the new Tucson Community Center—a glorified convention complex—opened in the middle of a gargantuan parking lot, where there was plenty of space for suburbanites to leave their cars when they came downtown to see the Ice Capades.

Tucson's experience with urban renewal was hardly unique. From the 1950s through the '70s, immigrant populations and communities of color in cities all over America were forced from their homes under the pretense that their neighborhoods were slums, all so that a project catering to the wealthy could be built in their place. In New York City, seven thousand Black and Puerto Rican families in San Juan Hill were uprooted to make way for Lincoln Center; in Atlanta's Summerhill, 3,800 families were displaced by a sports arena. As a boy in the 1920s, my paternal grandfather lived in a tenement in Boston's West End, a rowdy community of Italian, Jewish, and eastern European immigrants. Between 1958 and 1959, his old building, along with the rest of the West End, was flattened. Over the next decade, six luxury apartments sprouted in the forty-six-acre expanse of new plazas and gardens, hosting just two thousand residents, compared to the twelve thousand who once lived there. Every time I see those generic high-rises from the other side of the Charles River, a fist of anger and anguish clenches in my chest.

It's a feeling Lydia Otero knows well. The historian grew up in Barrio Kroeger Lane, a few miles south of La Calle. When I met them at a coffee shop directly opposite the bland edifice of the city's main library downtown, they remembered, "My father and my mother read the newspaper every day. We got the *Citizen*, and when they read about Pueblo Center my father would always throw the paper on the table. We were very attached to downtown." Otero moved to California shortly after the new convention center opened. "If I'd stayed here I just would've gotten used to it. But every time I came back to visit, it felt like something was wrong."

Though troubled by the destruction of La Calle, Otero told me they'd left Tucson for a more fundamental reason: "I was queer and needed to get away." They eventually found work in Los Angeles as an electrician, a day job that helped support their work organizing a group called Gay and Lesbian Latinos Unidos. In 1988, Otero was elected president of the group and featured on the cover of its newsletter, *UNIDAD*, wearing a white A-shirt and a curly black mullet.

After more than a decade in California, Otero started taking college classes. Once they got their degree, they were invited back home to pursue a PhD at the U of A. It was in the university archives that they found a collection of work by the journalist Alva Torres, who spent forty years as a columnist for the *Citizen* and, in the 1970s, campaigned for the preservation of La Plaza de la Mesilla, a historic gathering place for the city's Mexican American community. Otero's dissertation on Torres led to *La Calle*, which was published in 2010 and now serves as the definitive history of urban renewal in Tucson. Only since *La Calle*'s release has the city—which many residents like to call the Old Pueblo, when they're not expounding about their progressive politics—finally begun to reckon with the displacement caused by the enormous patch of scar tissue downtown.

Otero was so frank about the legacy of urban renewal in Tucson that it became immediately obvious why they have accrued a reputation as a civic gadfly, with little tolerance for the glad-handing of its booster class. Otero told me they initially received a lot of criticism for *La Calle*, even if its arguments have gradually become an accepted component of Tucson's story. Even as they leaned deeply into an overstuffed coffee shop armchair, Otero's inner fire was palpable. "When you're convincing people to visit or move to a city, you want to make it seem amenable to them. Sometimes brown faces are kept in the background in creating that friendly, welcoming experience. My mother was a maid and worked at the Holiday Inn, and I try to make a point that she was important to tourism, but people don't look at that."

Otero now lives on the west side of the river, in Menlo Park, near a recently opened office of the Caterpillar corporation. "If you go to the front of that building, there's a big scoop, bigger than me and you. A little sign that says: 'Historic, do not touch.' This was used to destroy homes! I walk my dog there, and every time I pass that I have to reflect." So much time has passed since the Pueblo Center buildings went up that demand is growing to preserve them. "It's brutalist construction, right? You can't tear them down!"

. . .

In an interview from the late 1980s, the architect Judith Chafee told the *Arizona Daily Star*, "My perception of what should be built in the desert stems from having grown up in the desert. I grew up going up and down arroyos and knowing where it was cool and where the breezes blew." Chafee's mother had trained as an anthropologist before the family moved to Tucson from Chicago in 1932, and the architect remembered "a lot of talk in my childhood home about traditional cultures here and respect for them."

Though she saw herself as a creature specific to the Sonoran, the *Star*'s Margo Hernandez described Chafee as the prototypical modernist architect—that is, if the prototype were female. "She speaks slowly and enunciates each syllable through her deep, throaty voice," Hernandez wrote, after noting the "ever-present More cigarette in her hand." She continued: "Her hair is cut straight and falls just above her shoulders. Her glasses, black and round, dominate her face."

Chafee was educated far afield: she attended high school back in Chicago, then college in Bennington, Vermont, before enrolling at the Yale School of Architecture in 1956. There, she studied under some of the brightest luminaries of the modernist movement in America, including Louis Kahn, Philip Johnson, and Paul Rudolph, the last of whom hired her to his firm after she graduated. Chafee went on to also work for Walter Gropius, Eero Saarinen, and Edward Larrabee Barnes, but she would later confess that their high-flying firms left her "very disheartened by the politics of New York architecture." Chafee had been the only woman in her class at Yale and was typically only one of a handful of female employees. She came to prefer drafting plans alone at her apartment in New Haven to commuting to Rudolph's office; when she was working for Barnes, she wrote him a letter expressing her displeasure with his staid approach to challenging conventions in the field, including the unspoken prohibition on elevating women into positions of leadership.

In 1969, Chafee, frustrated and homesick, abandoned mainstream American architecture once and for all. Reflecting on her last years on the East Coast for the journal *Artspace*, Chafee wrote that she would often flee along the suburban Connecticut road with the Volkswagen on a high whine. In the woods are perfect glass houses, in the villages priggishly perfect white churches. The sides of the road are perfect. Placed stones, considered fences, ground covers and orange day lilies. The trees are beautiful, they crowd over the road and touch above. I think that I will drown in the green muck. I think that I will lose myself, my purpose, if I can't see something beautifully naked and clear, if I do not see the edge of space—the horizon. I go to the seashore and find the space of the desert.

Chafee's words resonate with any southwesterner who has spent too much time on the East Coast. The only setting I've found that can approximate the sensation of vastness bestowed by cresting a ridge in the desert and then watching the titanic bowl of an arid valley extend out the other side is the Cape Cod National Seashore. A forty-mile arc of white sand, the National Seashore more clearly delimits the thresholds between earth, water, and sky than the rest of the region's island-spotted coastline. There, the collapse of dramatic cliffs into wide, windswept beaches being dragged under by the cresting waves of the North Atlantic recalls what my wife, Tess (a born and bred Bostonian), calls the "big nature" of the West. You can walk for an hour on the sand and be made small by the slow majesty of the geologic systems that created the Cape, operating on a timescale beyond human perception.

Necessary interior space is created when, faced with the immensity of Earth, one's ego is shrunken down to an appropriately miniscule size. Judith Chafee's decision to give up pursuit of the glamorous career that could've been hers had she stayed in the East was precipitated by the deep phoniness of the people and institutions that might've ushered her to fame. Surrounded by men who believed themselves to be giants, Chafee declined to play pretend. As she once put it to an interviewer, "The question is, 'Do you want to sell honest buildings, or do you want to be a movie star?'"

After coming home, the architect's first order of business was to set up an office for herself. She moved into four rowhouses in El Presidio, two of which dated to 1871, while the other two had been built sixty years later by the father of Monica Flinn, whose restaurant, El Charro, stood next door. The block was in rough shape: the adobe walls of the buildings were full of holes and only a few floorboards were scattered over the bare dirt inside. In one room, Chafee found a pile of garbage that went up to her waist. Undeterred, she combined the lots into a single complex that could serve as both office and residence, restoring the stucco, installing proper floors, and removing one of the original roofs to create an interior courtyard.

Today, the studio houses Poster Mirto McDonald, a firm cofounded by one of Chafee's early assistants, Corky Poster. It remains much as it was when Chafee retired in the 1990s. When I visited, a sliding door on one side of the studio's central room had been left open to let in a breeze from the courtyard, which was richly planted around a concrete basin that Chafee used as an outdoor shower but now held an enormous, branching spider plant. "My desk was over there forty-nine years ago," Poster said, pointing from his seat at the conference table to an adjacent room and chuckling. "I haven't made much progress."

Poster moved to Tucson the day after he got his degree from Harvard's Graduate School of Design in 1973. Chafee hired him entirely on the strength of that credential. When she interviewed Poster, Chafee was matter-of-fact: "I'm not looking at your portfolio, you can start on Monday." In his short time in the office, Poster worked on the plans for Chafee's most famous building, the Ramada House, situated in the foothills of the Santa Catalina Mountains. The structure takes its name from the shade structures first built by the ancestral Sonoran Desert peoples, but instead of human-scale structure made of saguaro ribs, Chafee's is a massive wooden latticework that floats over the house like a square halo, creating a play of rectilinear shadows on the white masonry of the house itself.

The architectural scholar William J. R. Curtis hailed the Ramada House for how it fused modernist massing and "Corbusian ideas such as the free plan, the grid of pilotis, and the shading slab" with "desert archetypes from the Native American tradition." It was the sort of building, he wrote, that "seemed able to draw upon indigenous wisdom, but without simply imitating vernacular forms: to penetrate beyond the obvious features of regional style to some deeper mythical structures rooted in past adjustments to landscape and climate."

Chafee would go on to build several dozen homes around Tucson, establishing a reputation for herself as one of Arizona's foremost architects even as she struggled to win commissions for the sort of massive public projects that expand an architect's reach beyond the handful of families rich enough to afford her. "She was uncompromising," Poster said, remembering a meeting with Jane and Peter Salomon, the couple who hired her to design what became the Ramada House. In the same room where I was sitting with Poster, Chafee showed the Salomons an estimate for the building's construction. They blanched. Looking over the plans, Peter realized that the ramada and the house were each essentially freestanding structures, and suggested, "Let's just get rid of the ramada?" Poster shook his head and grinned. "Judith gave him this *withering* look—I'm surprised he didn't disintegrate on the spot—and he said, 'Oh, well, I guess that's not a great idea.'"

Chafee may not have had an appetite for social climbing, but that didn't mean she lacked self-confidence. The architect's faith in her talent was unshakable, as was her attention to detail. "When she drew interior elevations," Poster said, "she would draw where the plugs were and dimension them so that everything was exactly where she wanted it. And I've learned since that if you *don't* do that, the electricians will put it cockeye in the wrong place." The soundness of this approach combined with her artistic vision left Chafee with little doubt about how her buildings stacked up locally, telling *The Arizona Republic*, "It's the best architecture in Tucson." Or, as her longtime assistant Kathryn McGuire put it: "We don't do junk."

Shortly after her studio was up and running, Chafee designed a home for her mother out on the far western flank of the city, where unpaved streets meander their way up a gentle grade until eventually meeting the sun-blasted Tucson Mountains. Though she had lived in Tucson for most of her life, Kathy McGuire still had trouble navigating this part of the city. After she picked me up on Speedway, she missed a turn in the foothills and ended up parking behind the house. To get to the front door, we had to skirt the empty pool on the structure's north side by foot, treading gingerly over the golden-flowering chamisa underbrush.

The white exterior of the Viewpoint House features three steps, creating a set of north-facing, clerestory windows that diffuse light throughout the single floor of the interior with a minimum of heat gain, which is likewise tamped down by the brise-soleils that jut out over the east-facing windows like hulking concrete eyebrows. We met the owner of the house, John Biklen, under one of these solar shades. "So this is called Viewpoint, and you can see the view," Biklen said, gesturing back toward the city, which unspooled before us on its way to meet the Santa Catalinas at the other end of the valley. When she was first planning the house, McGuire said, Chafee would spend hours at the site with her mother, bringing along a "picnic basket and martinis, just to get a sense of the wind and views."

Biklen bought the house with his partner, David Streeter, in 2004. They were collectors who, before retiring, ran an antiquarian bookstore in Southern California specializing in food and wine. The couple spent a few years in Palm Springs but couldn't afford any of that city's modernist homes, designed by the likes of Richard Neutra and Albert Frey. After their friend Laura Wills (the owner of Screaming Mimi's, a vintage clothing store in New York City) bought a house in Tucson, Biklen and Streeter visited and realized that the city boasted a similar but less recognized collection of modernist homes. When a listing for the Viewpoint House went online, John called David over to the computer and told him, "There's a Judith Chafee house for sale, and we can afford it." David sat down and clicked through the photos. Only a minute or so went by before he said, "John, we're going to buy that house."

As we entered, McGuire remarked that "every wall, every surface, was a different color" when Biklen and Streeter moved in, making the house a baffling kaleidoscopic of purple, orange, yellow, and what Biklen called "military green." All that's gone now: the walls have been restored to crisp white, the concrete that frames each window polished, black paint peeled off both the cabinetry and the metal ductwork that Chafee intended to remain exposed. The mechanical innards of the home were meant to be "like a trumpet," McGuire said, "announcing itself"—a tangible example of what Chafee meant when she said a building ought to be honest. Not only should its form be transparent, but also the systems that make that form habitable.

Despite all the concrete and galvanized steel, the interior of the Viewpoint House felt more therapeutic than industrial, all thanks to the clerestory windows, which gave the interior an ambient glow, neither bright nor dim. After the Viewpoint House was completed, it was featured in a *Los Angeles Times* article where Chafee explained her reasoning for all the measures that had been put in place to dampen the heat that including so many windows on a desert house would otherwise produce. Traditional, pueblo-style buildings, Chafee said, "were pleasant, cool caves to enter, but once inside these dark spaces, the glare from the punched windows is so intense that one's eyes struggle in constant flux, adjusting to the interior and to the view outside." Conversely, at the Viewpoint House, "all openings on the south and east are protected by overhangs that admit direct sun only for a short time in midwinter. There are no openings on the wretched west."

The brise-soleils, the careful attention to cardinal orientation—it all added up to a sophisticated system for separating light from heat. Which isn't to say Chafee was exclusively prioritizing function. "What's interesting," Biklen said at one point on our tour, "is she designed this window almost like a painting. It frames the mountains." McGuire added, "You always know where you are in her houses."

The careful balance between climate and grandeur was something Chafee continued to refine throughout her career. When I entered the Jacobson House in the foothills of the Santa Catalinas, I was stopped short by the home's library, which doubles as a staircase up to a reading nook, with each step holding a glass-fronted shelf of books. In a monograph about Chafee that McGuire wrote with the architectural historian Christopher

Domin, they suggested that the placement of this stepped library stair was a sly homage to the nineteenth-century architect H. H. Richardson, who often organized his grandiose Romanesque buildings around a central staircase, as in the New York State Capitol Building. In the Jacobson House, Chafee conjured the same drama but on a more intimate scale.

Demion Clinco is a former state representative who now leads the Tucson Historic Preservation Foundation, which puts on an annual Modernism Week to celebrate the city's postwar design heritage. Clinco had recently purchased the Jacobson House and was planning to restore it in order to rent it out to guests as a way to raise funds for the foundation. "I find the house to be really seductive. The way it interacts with the desert and the light—it's a place where there's no TV and you don't even really need internet."

Just off the library was the high-ceilinged studio where Joan Jacobson did her weaving (Clinco lovingly referred to this as the "loom room"). In the kitchen, Clinco showed me some boxes of Chafee's books that McGuire had recently dropped off, including a Bauhaus pamphlet in the original German and a French edition of a volume by Corbusier. We passed through the bedroom, with its personal, whitewashed fireplace, the visible flue lit elegantly by natural light, and then lingered in the living room, made invitingly bright by a pair of clerestories facing each other rather than the all-glass north wall, which framed a dramatic view of the emerald-tinted Santa Catalinas.

"The house almost acts as an aperture, sort of like a James Turrell," Clinco said, referring to the artist who has made this sort of framing of light and space his signature, whether in the eerie, blue tunnel of *The Light Inside* or his enormous land-art installation at the Roden Crater in northern Arizona. "The house almost seems to accelerate the speed of the sunset. Sit here with a few friends and a cocktail, the hyper-drama of it is so magnificent because it's framing out the sky and forcing your perspective. This beautiful concept of where you should be looking—that's where the house is oriented." The sense of awe in Clinco's voice reminded me of a remark Chafee had made about another one of her designs. "It's got a lot of spatial delight," she'd said. "I do think there's a sort of spiritual uplifting in a new architectural experience that makes people see things in a different way."

The house Lydia Otero grew up in, on the corner of Farmington Road and 24th Street in Barrio Kroeger Lane, had two views. From the side yard, Otero could admire Sentinel Peak, the 2,900-foot-tall hump of basalt better known as A Mountain, for the logo of the University of Arizona that decorates its summit. From the front porch, the vista was significantly less impressive: I-10, just two hundred feet from the front door.

Otero's parents had deep roots in Tucson. Their father, Daniel Otero, descended from a family that had received the first Spanish land grant for the area then known as Tubac in 1789, while their mother, Maria de la Cruz Robles, was the child of an Apache born in 1887—the year after Geronimo was captured—and a Tucsonenses who was at least half Seri, an Indigenous people from where the Sonoran Desert meets the Sea of Cortez. As with most non-Anglos in the Southwest, the arrival of American settlers destabilized both families: by the time he married, Daniel had been edged out of whatever inheritance might have been left to him by a cousin and was working as a desk clerk at the Hotel Apache downtown. Meanwhile, Chita—the nickname Maria went by—had dropped out of school to help support her nine siblings by working as a housekeeper. Though Chita's family had lived in Tucson for three generations, Otero writes in their memoir that Chita was so wary of having her right to live in the city questioned that she carried a U.S. Citizen Identification Card with her at all times.

Understanding that owning property was one of the few ways to guarantee some measure of stability for themselves, Daniel and Chita bought a large parcel of land in Barrio Kroeger Lane in 1940, even though they did not have enough money to construct a house. It was only a mile and a half south of downtown Tucson, but the city was then still small enough for their neighborhood to be considered the outskirts. Otero wrote that the area "must have seemed like paradise" to their mother, thanks to the proximity of the Santa Cruz River and all the farms along its banks. A Chinese immigrant named Howard Lee had also recently established a market nearby, and, down the way, "an enterprising gardener made a living selling zinnias, marigolds, and fresh corn."

The couple raised funds for their house by selling off some of their land; while Daniel fought overseas during World War II, Chita coordinated the efforts of ten family members to build a house out of adobe bricks. Everyone, Otero wrote, "lived on the construction site, sheltering under the ramadas they made. They dug a deep hole for an outhouse and a well for water, and they used the dirt from these projects to make adobe bricks in the large yard." Once the bricks had dried, they were assembled to form a room. Over the ensuing decades, as the family grew, five more rooms would gradually be added to the house.

"Have you ever made adobe?" asked Raquel Gutiérrez in an essay about the building material that has long been synonymous with the architecture of the Southwest. "Making adobe bricks is a process that requires repetitive motion of fingers, hands, wrists, arms, elbows, and shoulders as you're bent at both waist and knee," Gutiérrez wrote, before citing the process's origins in the tenth century. The "early constructors" would fill trenches "with a three-to-one mix of mud and broken stone, or river pebbles, tamping thoroughly to fill all of the nooks and crannies." After that foundation was "allowed to set from one full moon to the next," mesquite branches were used to create a scaffold that could be plastered the same way.

Whether adobe walls are constructed out of bricks or via this sort of wattle-and-daub technique, their utility is the same. Dried mud doesn't provide much insulation, but it does have excellent thermal capacity, meaning it can absorb and store heat over long periods of time. The industrial geologist George Austin described how this thermal capacity allows adobe to maintain stable interior temperatures with a sort of "flywheel effect": the material both absorbs heat from the sun during the day and also radiates warmth when nightfall causes temperatures to drop. As a result, the interior of an adobe structure never gets too cold or too hot.

Windows limit the thermal capacity of adobe, hence the dimness and glare that Judith Chafee found so challenging. For her part, Chafee did occasionally embrace the material, most notably in the Russell-Randolph Residence, which followed the ground plan of a classic Mexican courtyard home the owners had once encountered in Hermosillo. One of the few houses Chafee didn't design in the foothills, Russell-Randolph sits in a glade of mesquite and eucalyptus near the Santa Cruz River; panoramic views are less a concern than fostering a sense of comfort in the natural world. The editor Joseph Wilder wrote that, while he typically found all the raw concrete and metal in Chafee's houses "brutal, harsh, and cold," this one had an unmistakable warmth to it that he attributed to its embrace of tradition. "The materiality, the modern line combined with the adobe," he wrote. "I realized she had that language of modern southwestern architecture."

However humble, Lydia Otero's childhood home originally had a similar feel: in their memoir, the historian describes the fruit trees and vegetable garden their mother planted in the backyard, as well as "grapes whose vines attached themselves to trellises . . . and made for a cool and inviting outside space." But as more and more of the Santa Cruz Valley's groundwater was pumped to fuel Tucson's furious growth in the 1950s—over the course of that decade, the city more than quadrupled in size—the water table dropped and the Oteros' plants withered.

Meanwhile, the city bought up the 150 homes in Barrio Kroeger Lane that stood in the way of a proposed interstate. When it was completed in the mid-1950s, the freeway was at grade, meaning that kids from the barrios who lived on both sides of it would often dart across the lanes, dodging traffic so they could play in the wide median—the closest thing they had to a neighborhood park. Only a few years after the freeway opened, Otero's Apache grandfather, Pa' Luis Robles, was killed attempting to cross it. *The Tucson Citizen* reported that "several drivers ran over the body of Robles before the hit and run accident was discovered." Otero's sister watched from the front porch as the police stopped traffic to pick up the pieces of her grandfather's body.

The lethal cost of the city's neglect of neighborhoods like Barrio Kroeger Lane plagued the Otero family throughout their time on Farmington Road. Like many other neighborhoods of color whose segregation from the surrounding city was reified by the imposition of a highway—West Baltimore, Southeast Los Angeles, Houston's Fifth Ward—the traditionally Hispanic and Indigenous south side of Tucson was treated as an environmental and municipal hazard. Most of the gravel pits that provided raw materials for making the concrete used to build Tucson's subdivisions were located there; despite repeated requests from neighborhood associations, the companies that maintained the facilities refused to fence them, even though some included a pool of water as large as a football field and several feet deep. In 1956, Otero's nine-year-old brother, Jose Luis, drowned while playing in one of the gravel pits. Though one boy had died the same way a few years earlier, and another boy, named Jose Tapia, would die two years later in the same pit that killed Jose Luis, children were still drowning in the pits as late as 1969, when seven-year-old Robert Sepulveda died. The affected families received little relief from the negligent companies: the Oteros and Jose Tapia's parents both sued, but each only received a settlement of $4,500.

When it came to even seemingly benign urban planning decisions about how to deal with stormwater, the city simply pretended nobody actually lived on the south side. When the Davis-Monthan Air Force Base was expanded, a flood diversion channel was built to prevent monsoon rains from flooding the facility's new runways. That diversion channel ran to the Santa Cruz River via an arroyo behind the Oteros' house that was not nearly deep enough to accommodate all the extra water. When the next summer monsoon arrived, the entire neighborhood flooded. Otero wrote that the house's concrete floors and the bricks that propped up its couches and chairs were not "design choices. Until I was about nine, when it rained heavily, water ran into and throughout the house . . . I recall once standing on the sofa and rooting for a long thin snake that tried to fight the current before it too got swept outside."

In 2015, Otero wrote an op-ed for the *Tucson Weekly* tracing a through line between the demolition of La Calle, the deprivation of neighborhoods like Kroeger Lane, and the city's promotion of its mid-century architecture. "The Modernistic era symbolizes a period when, more so than today, powerful real estate brokers and developers dictated the planning agenda of Tucson and

Pima County," Otero wrote. "The same forces that encouraged growth and had the most to gain in the Modernist period fought fiercely to defeat public housing referendums that would have provided for the poor," while at the same time competing for federal urban renewal funds. "Despite some great individual designs in Tucson, Modernistic impulses deprived the urban core of vitality for more than forty years. The sprawl it produced made the city overly dependent on automobiles and fossil fuels. Those left behind during the Modernist period are still trying to catch up."

Though I admired both the aesthetics of Judith Chafee's architecture and the thoughtful way she fused modernist and traditional forms, the more of her houses I saw the harder it was to shake the sense of them as hovering above and apart from the desperate circumstances and institutional neglect that so many Tucsonenses suffered at the same time they were built. Those misgivings were only exacerbated by the fact that many of her clients were wealthy snowbirds who only ever used the houses in the wintertime, the season when Chafee's laudable efforts to tame the heat became superfluous.

McGuire and I said our goodbyes to John Biklen and headed across town to the Santa Catalinas. The rains of the summer monsoon had left the underbrush lush and vibrant, while the saguaros were practically bursting with all the moisture they'd absorbed. We stopped by the Ramada House, which had recently changed hands between Jane London, who had lived there since divorcing Peter Salomon shortly after the residence was completed, and a Hollywood couple: Sharon Jackson, an agent who once repped the likes of Jonah Hill and Elisabeth Moss, and her husband, Woody, a composer of soundtracks for TV shows and video games.

Since the Jacksons were in the middle of restoring the landmark, we couldn't go inside, but McGuire found a spot on a side street where there was enough of a gap between the crooked branches of a thicket of paloverde trees to see the white walls of the house under the yellowish wood of the new ramada, which had been installed so recently that the sallow color of the lumber was only beginning to fade. "There we are!" McGuire said as we pulled up. "And the right time of day"—since the sun was at a perfect angle to show off the shadows cast by the ramada on the house.

Though McGuire is typically reserved, she became buoyant as she described the Ramada House, repeatedly apologizing that we couldn't go inside. I was happy just to gaze at the house on its hilltop, as if it were the subject of a photorealistic portrait in a gallery. I pointed out how luminous the exterior walls looked under the high sun. "And they're not *white* white," McGuire replied, dreamily. "The sun's hitting them, but you can kind of see a grayness. See the foreground one is a little grayer? Really, they're the same color, it's just how the sun is hitting it." We rolled away but paused again when the Ramada House came back into view. "Ah, I love that house!" McGuire exclaimed. "When you get the shade on there it just . . ." she trailed off into a pleasurable sigh.

McGuire began working in Chafee's office in 1980, around the time the older architect had moved from her office in El Presidio to a townhouse near the Rillito River and converted her previous living quarters into more studio space. Though she had become well established in the city, Chafee was still routinely disrespected by clients who couldn't seem to fathom the concept of a woman architect. McGuire remembered one meeting that Chafee and another of her assistants, Robert Earl, took in the studio with a representative from the Tucson public school system about a remodeling project. "He and Judith were sitting at one end of the table and someone from the school was sitting on the other side. The whole time the client was talking, they were talking to Robert." Earl inclined his head toward Chafee, pleading with his body language, "you need to talk to her." McGuire shook her head. "It wasn't strange in those days." When I brought up the lack of any big public commissions in Chafee's portfolio that might've expanded her practice beyond second homes for wealthy outsiders, McGuire turned downcast. "We tried. We did try. And we would come in second. It kind of had that feeling: If you weren't a woman, we would probably give it to you." If Chafee came off as caustic or aloof in her later years, no wonder. It didn't seem to matter how talented she was, local powerbrokers had pegged her as a certain type of architect, and that was the only work she would ever be able to pursue.

We took Campbell Avenue up into the farthest reaches of the Santa Catalina foothills, passing through the entry gate for the Skyline Country Club Estates and by a mansion that was rumored to have been built for Oliver North and another that McGuire thought had recently been bought by someone with ties to Saudi Arabia. We kept going north until we reached the invisible line where the city limits butt up against National Forest Service land, a green-swathed Mount Kimball rising imperiously above us. There, as high up in the foothills as you can drive, sits the Rieveschl House.

Completed in 1988, the house is named for its first owners: George Rieveschl, the chemist who invented Benadryl, and his wife, Rose. Rather than a conventional hilltop site, the Rieveschl House is built in the shoulder between two hills. The west side holds guest accommodations that are connected to the main wing by a long hallway running between the two hills. If seen from overhead, the house looks like a barbell. In her notes, Chafee wrote that, rather than focus solely on light with this residence, the design had "the philosophical goals of disturbing the ground as little as possible and of living directly on the site determined by the structure." That meant supporting the bulk of the house, which cantilevers out over the steeply descending hillside, with a grid of thin columns and contorting its shape to avoid disturbing any of the nineteen saguaros on the site.

After we parked in front of the garage on a rivulet of water running down from the mountainside, McGuire led me through the sliding front gate to an exterior courtyard, where a wall of ruddy granite was framed by a concrete awning, the spiny fronds of an ocotillo at its center. The housekeeper, a chummy woman named Mara, greeted us in a Harley-Davidson T-shirt. She showed off a citrine geode that had recently tumbled down from Mount Kimball, then ushered us through a front door made entirely of stainless steel and glass. Mara flicked on the lights of the long corridor that connected the guest quarters to the main house and we followed the reflections of the orange rope light that ran along the hallway's polished concrete floor. The current owners, Sandra Helton and Norm Edelson, were spending the summer at their home in Chicago, but Mara had plenty of upkeep to do on the house, particularly when it came to ensuring that rain didn't leach into the building. As McGuire guided me into the main living area of the house, Mara rolled up the cord of a vacuum cleaner and turned down the classic rock she'd been listening to on the radio.

In the music's absence, our voices resonated across the vast space of the 7,200-square-foot house. After McGuire pointed out a jigsawing wall that had been tailored around an enormous saguaro and explained that all the niches in the living room had been specifically sized for the Rieveschls' collection of statuary, I was drawn outside to the mansion's balcony, which extends off the cavernous living room to the point on the hillside that was as far as the concrete mixer's pump apparatus could extend.

On the balcony, a steady wind took the edge off the intense heat of the day and my eyes rested on the city below, the tips of saguaros poking out of the paloverdes in the foreground, the Tucson Mountains a silhouette of roughened lapis in the distance. Everywhere in between, the checkerboard of beige rooftops and greenery rendered the city into a vision of desert abundance. For all intents and purposes, the south side was invisible. For anyone who could afford a house like this, I imagined, it would be hard to accept Ofelia Zepeda's suggestion: "We're not living right." On the other hand, for a kid in the barrio watching freeway traffic from their front door, the statement was so obviously true it would hardly be worth articulating.

Looking out over the valley, my mind leapt to the Getty Center in Los Angeles, Richard Meier's modernist palace with its divine perch in the Santa Monica Mountains. Though ostensibly a museum, the Getty is more memorable as the only overlook in that hulking city where you actually get to clock the Pacific and imagine its vastness swallowing up the works of humankind. Though the honest concrete of Chafee's balcony lacked the flair of Meier's fanciful Italian travertine, the feeling it inspired was the same. Standing out in the wind—as warm as the sea breeze is cool—the Sonoran seemed, for a second, like an ocean. Every house it contained, a lifeboat.

As much as the financiers, technologists, and career soldiers of Dallas, Austin, and San Antonio sometimes like to think of themselves as inhabiting the Southwest, Texas only really dries out once you cross the 100th meridian, near Abilene. Then it's about Colorado City where you get into the hundred-mile West Texas dereliction zone, where metronymic pumpjacks and drilling rigs are what passes for company in the manmade no-man's-land surrounding Big Spring, Odessa, and Midland. Only after I-20 merges with I-10 and the brown earth starts to roll into the hills that flank the Guadalupe Mountains can you be sure you've left the rest of the Lone Star State behind and entered the unlikely cradle of the great Chihuahuan Desert.

For generations, merchants from El Paso del Norte would make the two-day, seventy-mile trek to the salt flats just west of the Guadalupes, gathering salt there for distribution throughout Chihuahua and Sonora. After the Treaty of Guadalupe Hidalgo, Anglo efforts to privatize the salt flats became a political flashpoint, culminating in 1877, when Charles Howard filed claim to the land. The Mexican community revolted when Howard tried to arrest two men for trespassing on the flats, holding the businessman captive in the farm town of San Elizario.

After Howard was freed, he tracked down the state legislator Louis Cardis, whom he blamed for popular resistance to the scheme. Howard cornered Cardis on San Francisco Street in El Paso, blasting him in the chest with a shotgun. Fearing retribution for the murder, Howard took shelter with the Texas Rangers, but his coterie was soon overpowered by an angry mob. Howard and two of the lawmen were executed by firing squad; afterward, a gang of vengeful Anglo vigilantes from Silver City, New Mexico, pillaged San Elizario, killing at least four Mexican farmers. The so-called Salt War only ended after the military arrived and many of the rioters fled to Mexico. In the end, the salt flats became just another mineral deposit to feed the profits of extractive industry.

I kept driving west. A little after Van Horn, one of the eastbound lanes of the freeway was closed and traffic was backed up for miles. Nearly every vehicle was a semitruck that had come in from Mexico, each one waiting its turn to zipper into the one remaining lane and then hit the gas, their drivers anxiously checking the clock and doing the mental math to figure out just how much time they needed to make up to get to their destination on time, somewhere back east where their load would feed the inexhaustible American hunger for new clothing, televisions, and cars.

Sun blasted through the windshield and the interstate curved through the pass between the bone-white peak of Sierra Blanca and the tawny Quitman Mountains, then began its gentle descent into the valley of the Rio Grande. The road came within a few miles of the river before tipping north to parallel a fifty-mile corridor of farmland. Chamisa and Russian thistle coated the hills, but it had been so long since the bushes got rain that they had turned nearly white. Still, down in the valley, there it was, a green sward of life clinging to the river, extending all the way to San Elizario before finally giving way to the industrial parks, freight lots, and border checkpoints of El Paso del Norte.

CHIMERA

In a landscape where nothing officially exists (otherwise it would not be "desert"), absolutely anything becomes thinkable, and may consequently happen.

—RAYNER BANHAM

Sunset Heights sits on a bluff above downtown El Paso. The neighborhood's highest hill is crowned by a pump station that was built in 1910 to supply water to the rest of the city back when just thirty-nine thousand people lived there. At the time, that was enough to earn El Paso the title of largest city in the Southwest but still small enough that a well-placed water tank was sufficient to meet its demands. More than a century later, the building's exposed concrete foundation has been decorated with a mural depicting an even older water-delivery technology: two agua-dores transporting barrels on a mule-drawn cart.

I noticed the pump station while searching for an overlook in the neighborhood that might help me take in the full sweep of El Paso and Juárez, twin cities whose existence seemed forever shaped by the priorities of politicians and businessmen with little interest in ever actually visiting the borderlands. The best view I found was in Caruso Park, a strip of grass with two benches and a single pine tree. The June heat was beginning to rise as I gazed over the flaxen buildings that filled the valley of the Rio Grande; there, piercing the urban fabric like a long, implacable blade, was the border wall, its razor wire crown glittering in the early-morning sun. Further in the distance, a message in giant white letters was written across the face of one peak in the Juárez Mountains:

CD JUÁREZ
LA BIBLIA ES
LA VERDAD
LEELA

To the east, I could make out the bright white of the cross at the top of Mount Cristo Rey. Despite the oceanic political boundary that divided them, the two monuments announced a shared faith and culture that no government could ever hope to sever.

Only about five miles separate Juárez's Cerro Bola from the Franklin Mountains, the spare limestone ridge that jabs into the river's belly. It was this riverine gap that provided salvation to Don Juan de Oñate's settlers when they reached it in April 1598, and where the conquistador made his claim to the entire watershed of the Rio Grande: "I take possession, once, twice, and thrice, and all the times I can and must, of the actual jurisdiction, civil as well as criminal, of the lands of the said Rio del Norte, without exception whatsoever, with all its meadows and pasture grounds and passes." With that, Oñate nailed a cross to the nearest cottonwood tree and a group of soldiers fired a celebratory volley.

Oñate's declaration of control over a land he'd never seen was the first time El Paso del Norte contained a single man's ambition toward domination, but it would hardly be the last. Like all the grandiloquent resolutions that followed, Oñate's was as heedless about the practicalities of geography as it was about the countervailing desires of any and all people who might be unwittingly swept up in his vision. Reflecting on how precipitously the fortunes of El Paso fell over the course of the twentieth century— from the Southwest's preeminent commercial and industrial center to the region's poorest big city—I couldn't help but wonder if it was the delusions of grandeur of its civic leaders, each of them a little Oñate, that had made them so ill-equipped to navigate the economic and geopolitical shocks that have reverberated across the borderlands in recent decades.

After Oñate led his wagon train north to establish the colony of Nuevo Mexico, no Spaniard remained behind at the provisionary chapel that marked El Paso del Norte. The first permanent mission in the area—with the stated purpose of Christianizing the seminomadic Manso Indians who lived in proximity to the pass—wasn't established until 1659, and few Europeans lived in the region until the Pueblo Revolt in 1680, when the churches on either side of the river became a refuge for sixteen hundred fleeing Spanish settlers and the three hundred Indians who were allied with them, most of them drawn from either Isleta or the Piro speakers of the southern pueblos.

After Diego de Vargas's reconquest of Nuevo Mexico, El Paso del Norte remained a key waystation on the Camino Real de Tierra Adentro, the sixteen-hundred-mile trade route between Santa Fe and Mexico City. And it was thanks to the Camino Real that Mount Cristo Rey, the lone peak that stands in the narrow valley between the Cerro Bola and the Franklin Mountains, got its original name: Cerro de los Muleros, meaning Mule Drivers Mountain. The landmark was formed not by the grinding plate tectonics that upthrust the Rockies and the Sierra Madre but instead by an adventurous surge of magma that pressed the ground outward but never erupted, eventually cooling into a singular andesite peak. That made it an excellent point of navigation for mule drivers to track their progress through the vineyards that spread along the banks of the Rio Grande during the eighteenth century.

Like the Nuevomexicano settlements upriver, El Paso del Norte remained a sleepy agricultural area through the Mexican Revolution and the birth of the Texas Republic. The Treaty of Guadalupe Hidalgo, however, completely reordered the region by refashioning the Rio Grande as the border between the United States and Mexico. Overnight, the town of El Paso del Norte on the south side of the river was cut off from the much smaller villages on the opposite bank. Recognizing the commercial potential implicit in the new border, American settlers rushed in: midway through 1849, as many as fifteen hundred wagons were parked on the north bank of the Rio Grande. Those initial settlers gradually coalesced into an Anglo version of El Paso, just across the river from the city in Mexico that bore the same name—a Janus-like arrangement that presaged the formation of New and Old Albuquerque, albeit one in which the historic origin site was not in danger of being subsumed by its imitator.

After the railroad arrived in 1881, El Paso became the primary port of entry from northern Mexico, a hub for the importation of lumber and minerals, extractive industries that spread across Chihuahua in the late nineteenth century but were almost exclusively managed by American businessmen. Few profited more than the Guggenheim family, which was personally invited by the dictator Porfirio Díaz to take their pick of mining claims without having to pay any bothersome taxes. The Guggenheims subsequently acquired the American Smelting and Refining Company, or ASARCO, which operated the massive El Paso Smelter that processed much of the ore mined in northern Mexico so that it could be shipped to factories elsewhere in the United States.

The development of formal industry around El Paso paralleled a shadow economy: the city was becoming a redoubt of the West's fading libertine tradition, a Sin City that predated the founding of Las Vegas by decades. In the 1880s, El Paso's five preeminent madams each paid the city ten dollars a month to guarantee the safe operation of their brothels on Utah Street (today's South Mesa). Two blocks away, adobe saloons like the Pony lined South El Paso Street, wooden barrels of beer stacked outside the door.

Together, vice and industry helped El Paso become the original economic capital of the Southwest. And as more and more carousers moved in, the smuggling trade swelled in the city on the other bank of the Rio Grande—which was also called El Paso until 1888, when it was renamed to honor the liberal Mexican reformer Benito Juárez. Illicit commerce was facilitated by Mexico's establishment of an early free trade zone along much of its northern frontier, which allowed for goods to be stored there indefinitely without being subject to any duty, regardless of which side of the border would ultimately receive them.

In 1905, when El Paso outlawed prostitution, sex workers moved to Juárez; when Prohibition was ratified in 1919, saloon owners did the same. Hundreds of American bootleggers descended on El Paso, and after the bloodletting of the Mexican Revolution finally abated in 1920, the number of tourists crossing the Rio Grande soared to more than four hundred thousand, enough traffic to prompt the construction of two international bridges. Despite their professed temperance, El Paso's boosters excitedly promulgated the idea of their city as a gateway to Juárez, footing the bill for a national publicity campaign that reached 50 million people. The first Hilton Hotel opened in 1930 as an attempt to capitalize on the city's salacious allure. "Juárez is our greatest asset," said Bertram Orndorff, one of the hotel's funders. "High class tourists from the East on their way to California stop just to see Juárez. They have money to spend and they want to find a good time our neighbor across the river can give them—a dinner with liquor and other amusements." Texas maintained a ban on hard liquor even after Prohibition ended, ensuring that nightlife remained Juárez's primary industry for decades.

For El Pasoans coming of age in the 1940s and '50s, partying on the south bank of the Rio Grande was a rite of passage. The muralist and author José Antonio Burciaga wrote of how "high school football games and proms often ended in Juárez, where puberty, a fake ID card and money could get you into any bar." He became well acquainted with all the famous clubs that lined Avenida Juárez on the other side of the Santa Fe Bridge: the San Luis, where mariachis played corridos every night; El Lobby, a strip joint "where the meanest dancers boogied through the '40s, rhythm-and-blued in the '50s and rocked through the '60s and '70s to 'made in the USA' bands"; and the Noa Noa, where Juan Gabriel got his start.

For the better part of a century, El Paso del Norte remained a fluid point of intermingling between the peoples of America and Mexico. It was a place of shared culture and economic symbiosis, where the region's reputation as a debauched destination belied the deep Catholic faith that refigured El Cerro de los Muleros into Mount Cristo Rey. Gradually, though, the division between El Paso and Juárez calcified from a technicality to a chasm. Perhaps savvier leaders could have navigated that transition in a way that proved mutually beneficial to each city. Unfortunately, El Paso found itself governed by a gaggle of little Oñates, and it was their arrogance and egocentrism that ultimately prevailed over the common good.

When I met Bob Moore for coffee near the campus of the University of Texas at El Paso, he described how the antics of the city's politicians in the mid-twentieth century had provided cover for the city's decline. Moore has a gray crew cut and a sardonic smile—he's as close to a classical newsman as you're likely to find in the Southwest. With the exception of a short stint in Colorado, he'd been covering El Paso and southern New Mexico since 1986, a career that traced the contraction of the newspaper business from the boom times of the '80s to the devastation of the Great Recession, when advertising rates nose-dived and papers around the country began hemorrhaging print subscribers.

Moore led the *El Paso Times* from 2011 until 2017, when, he deadpanned, "I got tired of laying people off, so I laid myself off." After two years of freelancing for national outlets like *The Washington Post*, "I was really worried about the decline in local news, so I decided to start *El Paso Matters* and see what we could do." By 2022, the nonprofit website had a staff of eight journalists, their salaries covered mostly by grants from outside organizations. "The support has exceeded our expectations," Moore said, "but the long-term prospects are really uncertain. We have to be able to raise donations locally—in a low-income community like El Paso I'm not sure how that works."

Throughout the 1990s, El Paso's unemployment rate was two or three times the national average. Today, one in four children in the city live in poverty and the average El Pasoan makes about $17,000 less per year than other Americans. This is a far cry from the early twentieth century, when El Paso was not only a major tourist destination but also an industrial hub, a vital nexus of rail lines that connected Southern California and northern Mexico to the midsection of the United States. Moore pegs the inflection point for El Paso's decline to the late 1940s. "Coming out of

World War II, El Paso decided that its economic differentiator was going to be low-wage manufacturing," he said, a step away from both vice and industry that nevertheless still capitalized on the proximity of the border. Companies like Farah and Tony Lama Boots hired most of their labor force from Juárez, allowing them to pay cheap wages that kept prices for their garments low and profits high. Soon, El Paso had become the third-largest center of garment manufacturing in the country, after New York and Los Angeles.

That began to change in the 1960s, when Lyndon Johnson's administration canceled the Bracero Program, a policy dating to World War II that provided work visas to Mexican nationals with the assumption that they would fill jobs temporarily vacated by soldiers sent overseas. Most of the workers forced to return to Mexico remained in the borderlands, leading to unemployment rates of between 40 and 50 percent in Juárez, Tijuana, and Mexicali. In response, Mexico's federal government instituted the Programa de Industrialización Fronteriza, which created an Export Processing Zone, a newfangled Zona Libre where the government would waive duties on international companies that imported raw materials to assembly plants in the borderlands, factories known as maquiladoras. The first of these complexes, the Antonio J. Bermúdez Industrial Park in northeast Juárez, was completed in 1968. Over the course of the next decade, the Bermúdez facility grew to five hundred acres.

Meanwhile, Jonathan Rogers emerged as one of El Paso's most prominent businessmen. A Connecticut Yankee who first arrived at Fort Bliss as a six-foot-two Army lieutenant in 1950, Rogers married into the city's prominent Murchison family and took over its Mortgage Investment Company in the '60s after his father-in-law died. Rogers grew the company into a billion-dollar concern over the next two decades, and then, after a property tax increase angered his clients, decided that El Paso needed a CEO.

In 1981, Rogers campaigned on bringing a "business-like atmosphere" to the city's affairs, but once elected as mayor he developed a reputation as what the *El Paso Times* called a "tyrant-tycoon." He repeatedly tangled with the president of the police association, John Guerrero, over requests for more funding, including at one meeting when he interrupted Guerrero to say, "I speak when I decide to speak." When Guerrero replied by citing Robert's Rules of Order, the mayor shot him down again: "I follow Rogers' Rules."

To soften his image as a strongman, Rogers wore an orange guayabera to the city hall during his first summer in charge, declaring that "the guayabera is a part of our culture" in a nod to the city's enduring Mexican majority. In 1985, Rogers mandated that city officials stop wearing coats and ties entirely. *Time* magazine ran a picture of Rogers (this time in a white guayabera) grinning for the camera while he held a pair of scissors up to the red tie of a local banker. "A politician promises to cut back taxes or cut back the budget," the caption went. "But a mayor who promises to cut back neckties, him you had better take seriously."

Policy-wise, Rogers's administration was oriented around trimming the city's budget as much as possible. He refused to allow property taxes to budge, even as city services suffered. When the bus system needed an infusion of money, it was raised through a retrogressive sales tax. Meanwhile, Rogers's counterpart in Juárez, Francisco Barrio Terrazas, fired four hundred supposedly crooked police officers and raised the pay of the city's clerical workers to try to incentivize them to stop taking bribes. Those measures had little effect on corruption, but they did give cover for more than a hundred American companies to open plants in Juárez's ever-expanding industrial parks. In many cases, those shops replaced manufacturing facilities in El Paso.

Jonathan Rogers was tranquil about the changing economic landscape of the borderlands. "What problems?" he asked *Texas Monthly* when the magazine wrote about El Paso's faltering economy in 1987. "We're losing low-cost jobs and gaining high-cost," he said, as if the thousand people who had recently been laid off by Farah were outweighed by the three hundred new hires at the engineering firm Honeywell. "We will have an advanced technology center in two or three years. Meanwhile, we'll bring in more industry."

Since many laborers in Juárez spent as much as half of their income at shops in downtown El Paso, many business leaders on the other side of the Rio Grande were content that the two city's fortunes would rise in tandem. As an executive named Robert Head put it, "The more jobs there are in Juárez, the more pesos are put into El Paso. Anything that increases the attractiveness of either city will help both sides." By the late 1980s, it was clear that the benefits were in fact skewing south. One hundred thousand Juárenses were employed at a maquiladora, while the "twin plant" counterparts across the Rio Grande that completed assembly of the products made in Mexico (the final stage in the circumvention of typical import duties) had put vanishingly few people to work in El Paso, where the unemployment rate hit 13 percent.

Different leaders might have recognized that El Paso was in real trouble in the '80s. Unfortunately, the political class was riven by personal enmity and dysfunction, as exemplified by the election of Jay J. Armes to the city council. A decade earlier, Armes had become as close to a national celebrity as El Paso could boast, partially thanks to an appearance on Geraldo Rivera's after-hours talk show, *Good Night America*.

"Over the years, television and movie audiences have been flooded with imitations of James Bond," Rivera said by way of introduction. "In my opinion," he continued, "the person who comes closest in real life is private investigator Jay Armes. And his achievement is even more impressive because Jay Armes has established himself as a world-famous detective despite the fact that he lost his hands in a freak accident at the age of twelve." While Rivera narrated, footage rolled of a dim restaurant where a man in a white suit used hook-shaped metal prosthetics to wipe his mouth, before being accosted by two assailants whom he easily threw to the ground with those same hooks.

Julian Armas was born in Ysleta, the town originally founded by Christianized members of New Mexico's Isleta Pueblo who had fled the Pueblo Revolt. The accident that destroyed the boy's hands happened in 1946, when he and a friend were playing with a box of railroad torpedoes, the miniature explosives used to signal the need for an emergency stop to a train conductor. As his father, who worked as a butcher, relayed to the *El Paso Times*, "Julian said he had a torpedo cupped in the palm of his hand. He put his hands together, and began rubbing, with the torpedo between them. Then the explosion occurred." In a less innocent version of the story, Armas stabbed the incendiary with an ice pick.

After Armas became a newspaper boy for the *Times*, the paper wrote a feel-good follow-up story two years after the accident, printing a portrait of him on the front page with a tied bundle of newsprint dangling from his right hook. The reporter Cecilia Napoles wrote that Armas had learned to do just about everything with his hooks: he could "climb trees, comb his hair, shoot a gun," and even "played end and guard on his grammar school football team." Napoles wrote that Armas said he wanted to become a lawyer, though she also noted, "Theatrically inclined, he records little programs which he improvises himself. His hobbies are collecting coins and military souvenirs and reading detective books."

According to the memoir he published the year after appearing on *Good Night America*, Armas left for California after graduating from high school, where "I broke into the movies, although I never got to be a star." In his telling, Armas appeared in thirty-six movies and twenty-eight TV shows over the course of six years, enough work to finance an apartment in Beverly Hills and a Cadillac. When he wasn't on call, Armas capitalized on his "natural flair of the mimic" by studying German, Italian, French, and Chinese. He only returned to El Paso because "being a small-time movie actor wasn't the only thing I wanted to do with my life." Curiously, the stint in Los Angeles was completely absent from the story Armas told Geraldo Rivera. Wearing his white leisure suit and reclining on a soundstage in Hollywood, Jay Armes—which he referred to as "his real name" during the interview—said he enrolled in college at the age of fifteen, where he earned a degree in criminology.

It's impossible to say what the early years of Armes's career actually entailed, both before and after he established his PI shop, "The Investigators." What's clear is that he did make a few friends in the entertainment business. In 1972, he got a call from Marlon Brando, who asked him to recover his son Christian from Baja California, where he'd been taken by hippie associates of his mother, Anna Kashfi, then in the midst of a bitter divorce from

Brando. According to Armes, he rented a helicopter to execute the mission, personally piloting it all over the peninsula until he spotted a red VW bus hidden above a small bay near San Felipe. Armes landed his helicopter a mile away, then single-hook-edly rescued Christian from his eight supposed kidnappers, including a bearded man wielding a spear gun and a woman he discovered naked in her sleeping bag. "Don't worry, lady," Armes told her, "you're not my type."

An initial story about the rescue that the AP headlined "El Paso Investigator Finds Movie Stars' Son" triggered a raft of national coverage. That summer, the *Chicago Tribune*, *Detroit Free Press*, and *Miami Herald* all ran a feature about Armes by Ivor Davis. The legend grew in the years that followed, with *Newsweek* writing that Armes "keeps a loaded submachine gun in his $37,000 Rolls-Royce as protection against the next—and fourteenth—attempt on his life" and *People* naming him one of its "Twenty-Five Most Intriguing People in 1975." The many journalists who visited El Paso to meet Armes were all baffled by the juxtaposition of a real-life James Bond with his unprepossessing hometown. Davis described his "one million dollar, four-acre estate" in the city's economically depressed Lower Valley, where he kept a menagerie of safari animals in addition to a helicopter and a fleet of nine cars around a "colonial-modern, white-pillared home" that he'd transformed into a "virtual fortress" with a ten-foot-high fence and surveillance cameras.

In his memoir, which was ghostwritten by Frederick Nolan, Armes cited a long list of impressive clients, including the chairman of a San Francisco electronics company, a German diplomat, a New York City con artist, and the "Onion King of America." His retellings of the various "capers" they sent him on were full of the sort of overwrought detective schtick with which Armes had familiarized himself when he was fifteen, back when he was eagerly reading pulp novels in his downtime from either delivering newspapers or attending college criminology classes. "It gets so you hardly even blink when a strange woman calls you in the small hours of the morning and whispers desperately into the telephone that her husband is trying to kill her," he wrote, and related the con artist's confession: "I'm a fraud, Mr. Armes . . .

My whole life is like a charade played inside a nightmare. I'm thirty-two years old and everything is built on lies and deceit."

This swaggering image was a far cry from the soft-spoken man chatting with Geraldo on *Good Night America*. Though he may have worn an impeccably tailored suit and plastered his hair with mousse, Armes was decidedly awkward onscreen, narrating his accident and the rescue of Brando's son in strangely halting terms. Whenever Armes stammered or paused, Geraldo prodded him to continue, declining to ask any questions that might challenge the preposterous details of his stories.

When *Texas Monthly*'s Gary Cartwright visited Armes's home in 1976, he excused himself from a discussion of the detective's extensive gun collection to make a phone call. Through a porthole window in the house's bar, he could see the famous helicopter. "From appearances, it hadn't been off the ground in years," Cartwright wrote. "Its tires were deflated and hub-deep in hard ground, the blades were caked with dirt and grease, and the windows were covered with tape instead of glass. Armes had told us that the chopper had a brand-new engine. I wondered why he hadn't put glass in the windows."

Cartwright ably debunked the myths that Armes had clothed himself in, rejecting Geraldo's assertion that Armes's life was "a classic tale of truth being more exciting—or stranger—than fiction." He pointed out the numerous discrepancies in Armes's backstory, offering instead a childhood friend's remembrance of Armes returning from his brief stint in California in "an old, raggedly-topped Cadillac with a live lion in the back and a dummy telephone mounted to the dashboard." Though he lacked the resources to do the character he had created for himself justice, Armes had already reinvented himself as an international man of mystery.

To make good on that fantasy, Armes would need a benefactor. He seems to have found one in the grandson of Thomas Fortune Ryan, the Gilded Age tobacco and insurance magnate, whose estate was valued at $142 million when he died in 1928—more than $2 billion in twenty-first-century dollars. Thomas Fortune Ryan III cofounded Lockheed Aircraft and, in 1941, acquired the fabled Three Rivers Ranch near Tularosa, New Mexico.

According to Armes's memoir, Ryan walked into his office in El Paso looking to help a friend who was concerned about the welfare of his daughters. "Neatly, rather than elegantly dressed in western-style clothes, he was a good-looking man in his sixties with silver-gray hair. I tabbed Ryan as a middle-sized rancher, doing better than breakeven," Armes wrote. After receiving his retainer via a check that "was folded about ten different ways, and looked as if he'd been carrying it around since about 1901," Armes called a banking contact named Ken to check on the financial wherewithal of his newest client.

> I told him the name: Thomas F. Ryan. There was a funny sound at the other end of the phone, as if Ken was choking on a fishbone.
> "You really don't know who he is?"
> "No," I said. "I really don't know who he is."
> "Jay," Ken said, "Thomas Fortune Ryan the Third is one of the richest men in the United States, if not the world."

Even allowing for Armes's hyperbole, Ryan was the real deal—a scion of Eastern money seeking a bit of fresh air and privacy in the wilds of southern New Mexico. Armes went on to claim that he and Tom Ryan went into business together as owners of Radium Springs Projects, which purchased a 150-acre property near Las Cruces that "contains underground thermal springs holding water with a constant temperature of 190°C. Plans have been mooted for a thermal power plant on the site, although at the moment they are only at the drawing board stage."

While that scheme seems to have been another mirage (Radium Springs remains a mostly agricultural community along the Rio Grande), Gary Cartwright was told that Ryan had brought Armes in on some real estate investments. Aside from that, the journalist found evidence of a few potentially lucrative cases for Armes (namely the recovery of merchandise for a trucking company), but nothing that justified the million-dollar fees Armes claimed to charge.

Nevertheless, between the partnership with Ryan, the book,

his appearance as a villain named "Hookman" on an episode of *Hawaii Five-O*, and whatever local cases all that publicity generated for the Investigators, Armes surely earned enough money to live comfortably in the Lower Valley. "The real story," Cartwright wrote, "is of a Mexican-American kid from one of the most impoverished settlements in the United States, how he extracted himself from the wreckage of a crippling childhood accident and through the exercise of tenacity, courage, and wits became a moderately successful private investigator."

By the 1980s, Armes's national notoriety had faded, making him all the more intent on cementing his legend in El Paso. He made two unsuccessful runs for sheriff before trying for the city council in 1985, promising to "straighten out our Lower Valley." As he put it to a reporter from the *El Paso Herald-Post* in an interview at his "elaborately secure office" on Montana Street, "I'm a victim of the problems down there, just like the rest of the District Six people are victims. If we sit there and take it for another two years, it's the same thing as somebody kicking and kicking and kicking you and you stay there. What's going to happen? You're going to go into a permanent state of demise." He lost that race too, but managed to finally squeak into the city council four years later with only 140 votes to spare.

Armes's two years on the city council were among the most disruptive in memory—at one point, the city hall was evacuated because of a novelty clock on his desk that looked like a bomb. However muted his presence had been on Geraldo's show a decade earlier, Armes was an outspoken figure at city council meetings, particularly in his exchanges with a fellow councilor named Mateele Rittgers, whom he compared to a stand-up comedian and a cockroach. The name-calling came across as racist (Rittgers was Black), particularly after a hearing where Armes used the term "afro-engineered" to refer to poor craftsmanship. The head of the local NAACP, Johnnie Washington, was outraged, but when Bob Moore, then a cub reporter for the *El Paso Times*, called Armes to get a response to her comments, he claimed first that he hadn't used the term, and then, when Moore pointed out that the meeting was recorded, insisted that there wasn't any problem. "I just talked to Johnnie," Armes told

Moore, "and she's fine." When Moore reiterated that he had also just talked to Washington, Armes hung up the phone.

While Armes was hardly responsible for the bottom falling out of El Paso's economy, his preference for style over substance—shared by Jonathan Rogers and plenty of other city officials—helps to explain why the city was caught so off guard by the North American Free Trade Agreement. After the treaty came into effect in 1994, the number of maquiladoras in Juárez more than doubled and the garment industry in El Paso vaporized. At a Labor Day march organized by La Mujera Obrera, an organization that advocated for the women who made up 80 percent of El Paso's garment workers, one laborer told the researcher Francisca James Hernández that she had just been laid off after seven years at Johnson & Johnson. "They suddenly closed the factory and went to Juárez," the worker told Hernández. "We were like five hundred people . . . There will be more, many more." By 2000, Lee, Wrangler, and Levi's had all pulled out of El Paso, contributing to the six thousand garment workers in the city whose job was lost to NAFTA—the highest number in the United States.

Jonathan Rogers's promised transition to becoming an "advanced technology center" never happened: El Paso lacks either a top-flight federal research facility, like Albuquerque's Sandia National Laboratory, or the entrenched aerospace and technology industry of Phoenix. Even UTEP, with its twenty-four thousand students, is only half the size of the University of Arizona. The city's only equivalent lifeline was the massive Fort Bliss military complex, which supports the employment of more than 130,000 people—roughly a third of El Paso's labor force.

Few books better articulated the forlorn state of the city in the '90s than Dagoberto Gilb's novel *The Last Known Residence of Mickey Acuña*, which centers on the long stay of a drifter from New Mexico at the YMCA on Wyoming Avenue. As Mickey familiarizes himself with the neighborhood, which is cut off from downtown El Paso by I-10, he notices, "Beer bottles shattered at the edges of the sidewalks and in the gutters, glittering the dark alleys, colored an attitude that held the territory. Mickey didn't hear any hooves but felt they'd come beating up any minute."

He feels that he can see the Wild West of the city's past straining against its attempts to appear like any other contemporary
American metropole. "As honest as bleached blond hair on a dark
Mexican woman," Gilb wrote, "El Paso's truth was not beauty-
parlored well enough."

When I arrived at the base of Mount Cristo Rey in the summer
of 2022, the early-morning sun had already forced the thermom-
eter on my rental car up into the low 80s, even though it was not
yet 7 a.m. A sign in the dirt parking lot advised all visitors to alert
local police before they began hiking; I ignored that suggestion,
but did stick an ID—proof of citizenship—into my backpack
along with a granola bar and a full-size Nalgene of water.

That transformation of Cerro de los Muleros into a religious
site had begun in 1887, when ASARCO built a plant capable of
refining lead from mineral deposits in northern Mexico just a
few miles from the mountain. Most of the facility's laborers were
drawn from the company's mines in Chihuahua, and 2,500 of
them ended up living together on a portion of company land that
came to be known as Smeltertown.

Two years after the plant opened, a Spanish priest named
Lourdes Costa arrived in Smeltertown and set up a small church,
which he named San José de Cristo Rey. Costa had an aspirational streak, inflamed by a sense of vanity that led him to once
compare his own sharp-nosed profile to that of Franklin Delano
Roosevelt. Sitting at his desk in the church, he would often turn
his head to the window to gaze up the sheer slopes of Cerro de
los Muleros, imagining how to extend his domain. In the early
years of the Great Depression, the padre went to the bishop of
El Paso and told him he'd had a vision of placing a figure of Jesus
atop the mountain (a vision that just so happened to mirror the
Christ the Redeemer statue that was then being installed in Rio
de Janeiro). The bishop liked the idea so much that the Catholic Church purchased the two hundred acres that encompassed
Cerro de los Muleros.

In 1934, Costa's parishioners erected the first monument on
the peak of the renamed Mount Cristo Rey, a twelve-foot-tall

wooden cross. For the dedication, Costa climbed the mountain with his flock. Standing in front of the cross, the wind gusting his robes, Costa announced, "This spot is now consecrated and will become a place of pilgrimages in a short time, and the big multitudes will come from all parts of the country and Mexico." Within a few weeks, the wooden cross had been replaced by one made of iron that was forged at the plant in Smeltertown and engraved with a Latin rendering of the phrase the Roman Emperor Constantine supposedly saw hovering over the sun, prompting his conversion to Christianity: "In hoc signo vinces"—"In this sign thou shalt conquer."

Costa wasn't done. With the national economy still in tatters, he dipped deeper into the diocese's coffers to pay men from Smeltertown $1.50 a day to improve the road up to the top of Mount Cristo Rey and level its pinnacle so that a larger monument could be constructed. Then he enlisted Uribici Soler, a fellow Spaniard who had developed a reputation for the busts he sculpted of the Indigenous peoples of Latin America. Soler made plans for a forty-two-and-a-half-foot cross made of bright Cordovan limestone, which, rather than depicting Christ's martyrdom, would provide a vision of the savior after his triumphant rebirth.

The limestone for the statue was sourced from a quarry near Austin and shipped along the Southern Pacific railroad that ran through Smeltertown. Forty tons of stone were hauled up the mountain on a tractor in 1938, and Soler began work on the statue once the raw material was already in place. Having fully succumbed to Lourdes Costa's charisma, Soler continued work on the sculpture at his own expense even after the diocese of El Paso ran out of money. The Smeltertown parish dedicated the statue in 1939, and the mountain's transmutation into a pilgrimage site was given full fête a year later, when the bishop of El Paso held a five-hour mass at the summit in front of a crowd of twelve thousand people. So many pilgrims climbed the mountain from their homes in Texas, New Mexico, and Chihuahua that many ended up far downslope from the statue, digging their heels into the mountain's loose topsoil as they strained to hear the bishop's address over the wind.

After zipping my backpack, I found the route to the top of Cristo Rey under a metal arch where "Welcome Bienvenidios" was written, a white cross serving as keystone. No one was around when I started up the trail, which isn't to say I was alone: at the first bend in the path, I took a look back the way I had come and realized that a white truck with a distinctive green sash painted on its back door was parked on a rocky outcropping above the parking lot—Border Patrol, the fifth of their vehicles I'd seen that morning. Though the sense of being surveilled was chilling, I still couldn't shake a feeling of being propelled up the mountain—not by any sense of faith, but rather by the raw audacity of the place, by my desire to understand what would compel first one man and then an entire community to toil relentlessly at erecting a monument on this strange, lonely mountain.

On the right side of the path was a shrine to Saint Anthony, with a bilingual sign explaining that "some Christians pray to him to help them find lost objects." There was no chance of this Saint Anthony going missing, given that he was encased in a beige metal cage with weaving so tight it was nearly impossible to make out the figure inside. As I continued up the gently sloping path, I passed a similar shrine to Saint Joseph that had also been put under lock and key. I heard a whirring noise when I paused to drink some water and put sunscreen on my legs: a helicopter flying low through the valley.

Up ahead, just before the path began its ascent of the steepest part of the mountain via a half-dozen switchbacks, a narrower trail led west to another shrine. The same helicopter reared into view from behind the mountain, a green stripe on its tail. I walked over to the shrine and squinted through the mesh to see Our Lady of Fatima painted on tiles depicting a blue sky scattered with white clouds.

The helicopter circled again, its pilot clearly keeping an eye on me. Off to the south, I could see the border wall, which a colonia on the outskirts of Juárez called Anapra ran up against. The helicopter passed again, low now, its rotors shredding the quiet of the morning. I leaned down to pick up some litter left behind on the sandy apron in front of the blessed virgin. The label read "Electrolit: Suero Rehidratante." It was a Mexican version of Pedialyte, sabor manzana. Not much imagination needed to figure out how it got there. Once I was a little farther up the trail, I could see that the border wall stopped abruptly where the mountain began its rise out of the desert floor, creating an obvious circumvention point for migrants.

After throwing back a few more mouthfuls of water in a rare shady spot on the trail, I continued upward, stopping only briefly when two middle-aged men came trotting down from the summit. One, who looked so much like a Chicano Guy Fieri it startled me, looked grave, telling me, "Hey man, it's closed up there. It's a whole situation, they got it closed off." For a second, I believed him. But when his friend broke into a smile I grinned too, snapping back, "Oh yeah, they closed the whole cross?" He laughed and slapped me on the arm before continuing on his way.

After the last bend in the path, the wind gusted so hard I had to adjust my hat to keep it from blowing over into Mexico, just thirteen hundred feet away. A concrete ramp with a yellow handrail led up to the cross, which stood on a pediment studded with chipped tiles. In 1983, volunteers had organized a telethon that raised $30,000 to help realize Lourdes Costa's ultimate vision of surrounding Uribici Soler's Christ figure with a massive crown wrought in marble. Forty years on, it was far from the open-air cathedral the priest had dreamed up. My only company at the top were two teenagers who huddled together in the shade cast by the cross in a posture more lusty than prayerful. Over the rim of the crown, I saw the remains of some spotlights the volunteers had installed in the 1980s to illuminate the statue. In short order, all of the lighting equipment was stolen, even a generator encased in a concrete bunker. All that remained were a few crumbling plinths with rusted rebar jutting out the top.

Beyond, the view of the valley was stunning: only at a distance can you see El Paso and Juárez as the cohesive city they once were. The perilous border drawn artificially through the trunk of the Chihuahuan Desert was obvious on both sides of the mountain, yet completely absent on the formation itself. This gave Cristo Rey a centered feeling, marking not just the midpoint of El Paso and Juárez, but also that of the entire, two-thousand-mile extent of the borderlands. Looking out over the valley, the geopolitics were puerile. What really mattered was the way the people of El Paso del Norte leaned toward transcendence, an impulse common across the arid lands, where survival is synonymous with the creation of something out of nothing.

I searched the expanse for the Smeltertown cemetery, which still stands on a rocky hilltop sandwiched between two expressways, just over a mile east of the mountain. The cemetery has survived the community, which ASARCO bulldozed in 1973 after a researcher from the Centers for Disease Control found that 62 percent of the children living there were at severe risk of suffering permanent brain damage because they had elevated concentrations of lead in their blood, the consequence of living downwind of smokestacks that billowed hundreds of tons of the toxin every year. Those smokestacks had been built decades earlier to address persistent community complaints about pollution, as if it was the lack of sufficient elevation that was fouling the air. For all the prosperity that spread across America in the century following the Treaty of Guadalupe Hidalgo, it was vital to remember how many communities like Smeltertown suffered in order to enrich some faraway family like the Guggenheims, few of whom would deign to visit El Paso even as its laborers generated the capital that allowed them to build their museums in Manhattan, Venice, Abu Dhabi, and Bilbao.

In the short distance between Mount Cristo Rey's summit and the cemetery's weary patch of hallowed earth, the straining of the people of Smeltertown to overcome their circumstance and create something beautiful and lasting—a liminal space where it would be possible to both reflect on one's life and be overpowered by the sublime sensation of the manifold souls below—was palpable. The humility of that hope felt like an antidote to Lourdes Costa's self-glorification, even as I recognized that one urge could not exist without the other.

Gazing ahead into the cloudless sky was Uribici Soler's Christ of the Rockies. Even squinting up into the sun, I could appreciate the softness of Soler's rendering—the gentle folds of Jesus' robe, the easy drape of his belt, the relaxed tilt of his hands, the pacific expression on his face. This, again, was meant to be the risen Christ, victorious and eternal. If he really did look out over the borderlands from this fabulous vantage, how many kingdoms would he see?

PART II

EL PUEBLO UNIDO
JAMÁS SERÁ VENCIDO

Las Vegas–Albuquerque

DEFIANCE

I won't give up, I won't let up, and I won't shut up.

—MARZETTE LEWIS

The Kim Sisters' first performance on the Las Vegas Strip was a sensation. The evening of February 3, 1959, Sook-Ja, Min-Ja, and Ai-Ja entered the Thunderbird Hotel's lounge in matching pink taffeta dresses. As the crowd continued to chat among themselves, sipping martinis and swapping stories about fortunes won and lost on the green felt tables just outside the door, the three women began to harmonize the opening to "Sincerely," a love song that the McGuire Sisters had taken to the top of the Billboard Charts in 1955. The Kim Sisters had only been in the United States for two weeks. And only two of them were actually sisters. Nevertheless, by the time their set was over the initially inattentive audience at the Thunderbird roared its approval. In a review, the *Las Vegas Sun* wrote that "the best thing we can do is tear up their passports so the three little dolls can't return to South Korea."

Not that they were anxious to go home. Ai-Ja and Sook-Ja, who would soon begin going by "Sue," were the daughters of a famous couple, Kim Hae-Song, a composer, and Yi Nanyŏng, a singer known for 1935's sorrowful "Tears of Mokpo." The family had lived in a spacious home at the base of Seoul's Namsan Mountain before the war, but Hai-Song's outspoken opposition to communism made him a target during the North Korean occupation of the city; both he and Nanyŏng wound up in prison camps, forcing thirteen-year-old Sue to provide for her six siblings as a busker, singing on the street and begging for change. After Nanyŏng escaped and the family fled to an army base at the southern end of the Korean Peninsula, she taught the children American songs and booked them on the USO circuit. As Sue would later remember, from "1954 to 1958, we sang for the GI troops all the time. That's how we got to eat. Plain language, that's how we survived."

The children's success led Nanyŏng to formalize the group around Sue, Ai-Ja, and their cousin Min-Ja, and thanks to the Kim Sisters' growing popularity in Korea, an expat American talent manager was able to book them for a thirty-day run as members of the Thunderbird's China Doll Revue, an orientalist spectacular that included a pair of dancers named Dorothy Toy and Paul Wing, as well as the rockabilly group Ming and Ling—"the famous Chinese Hill-Billies." After the Kim Sisters landed at LAX, their booking agent, Tom Ball, picked them up in a black Cadillac. On the long drive to Las Vegas, the young women looked out over the Mojave, the low winter sun throwing the shadows of roadside Joshua trees into gnarled silhouettes. Sue leaned over to Ai-Ja and Min-Ja, her eyes wide. "I don't think this is America, do you?"

Sue Kim would become enmeshed with Las Vegas, just as the city itself was emerging into an unlikely but integral component of the nation's culture. Still, her observation continues to resonate, not just out on the eerie backroads of the Mojave but in the cacophonous, neon canyons of Las Vegas itself. Since the city first asserted its place in the national consciousness in the years after World War II, it has typically been understood as America's great outlier—more a marketing phenomenon than an actual place. As Tom Wolfe put it, "One can look at Las Vegas from a mile away on Route 91 and see no buildings, no trees, only signs." And, as the architects Denise Scott Brown and Robert Venturi added, "If you take the signs away, there is no place." The philosopher

Bruce Bégout called Las Vegas "a pretend city." "Las Vegas was born to radiate, to flash, to explode," he wrote, meaning that it was less an authentic location than "a mirthful and all-devouring abyss of energy."

There is no river running through the Las Vegas Valley that might have supported large-scale agriculture projects or connected the would-be city to commercial centers elsewhere. Consequently, the transformation of this arid anyplace into a global destination has always felt like a nifty trick. As if the whole city were a pop-up ad the country didn't mean to click on. It's a rare metropolis that was a sign and a signifier before it was an object. No wonder so many people claim to hate it even though they've never been.

For a southwesterner like me, though, Las Vegas feels like home. I first understood that while staying at an Airbnb in an apartment complex close to UNLV: a faded, two-level block with outdoor staircases and dirt landscaping—identical to the dilapidated building a friend of mine had moved into in Albuquerque after he dropped out of UNM. This same sensation of belonging returned on a walk down Charleston Boulevard. The constant rip of traffic on Charleston's six lanes, the scattering of Walgreens and cinder-block office buildings, the sun-blasted bus shelters: I might as well have been on Montgomery Boulevard in Albuquerque, on my way to the coffee shop I haunted as a teenager.

The image of Vegas that animates the imagination of millions of tourists from around the world is just that: an image. The city underneath it, however, is southwestern to the core. Only here, in the forlorn Mojave Desert, could a handful of civic pioneers have realized their supersize ambitions in such a short period of time. The city stands as an assertion of selfhood in a place given to abnegation, a monument to willfulness in the face of droughty indifference.

Though Las Vegas required the same collective effort to create as the older cities of the Southwest, its founders faced even more daunting ecological conditions yet built their metropolis in triple time. This frenzy initially perpetuated a society with defined winners and losers, the odds stacked in favor of anyone who fit the profile of the last person to hit a jackpot. The difference was that the everyday Las Vegans manning the tables decided that the newly established system wasn't going to work for them. Before it could calcify, they fought back.

From the origin of the ancestral Puebloan and Sonoran civilizations, around 500 C.E., to the arrival of the conquistadors a thousand years later and then the establishment of first European colonies and then American homesteads, the Las Vegas Valley remained largely uninhabited. Only on occasion would the nomadic Nuwuvi, also known as Southern Paiute, camp around its artesian springs for a few nights before moving on to more productive hunting grounds near the Colorado River. The first permanent settlement in the valley wasn't attempted until the 1850s, when a small group of Mormons was sent there by Brigham Young. They left after only a few years, unable to find any resources in the surrounding area valuable enough to justify staying, telling other members of the church it was a landscape the "Lord had forgotten."

The Mormons should have been looking farther north. It was in the hills between Carson City and Reno that silver was discovered in 1859, and so much wealth was extracted out of the Sierra Nevada in the ensuing months that Nevada was divided from the Utah Territory in 1861, then admitted to the war-torn Union as an independent state three years later. Virginia City, the boomtown nearest the Comstock Lode, swelled in the course of a decade to a population of twenty-five thousand people—almost half the residents of the entire state—with twenty-five theaters and more than a hundred saloons, before collapsing into a near ghost town once there was no more silver left to haul out of the earth.

Through all the boom and bust of the Silver State's early years, southern Nevada remained an afterthought. By the dawn of the twentieth century, only a few dozen homesteaders had found their way to the Las Vegas Valley, attempting to eke what living they could from the springs and meadows of fluff grass and red grama that gave the basin its name. Las Vegas itself would only be incorporated in 1905, after William Clark, whose copper mines in Montana had made him one of the wealthiest men in the West, built a station there for a railroad line connecting Salt Lake City to Los Angeles. His route ran straight through the valley so the local mines around Tonopah and Goldfield could send their minerals to market.

In 1931, the Bureau of Reclamation began plugging the Colorado River with a dam that would eventually be named after Herbert Hoover, the man who oversaw the plans to corral the waterway that delegates from the seven states it flowed through had agreed to at their conference in Santa Fe. The same year, Nevada legalized gambling: the legislature decided that it was better to regulate the industry than pretend backroom casinos weren't already thriving across the state.

"Conditions will be very little different than they are at the present time," wrote the editorial board of the *Las Vegas Evening Review-Journal*, "except that some things will be done openly that have previously been done in secret. The same resorts will do business in the same way, only somewhat more liberally and above-board." Nevertheless, organized crime figures took notice. With a new settlement, Boulder City, being founded to house the five thousand laborers erecting the Hoover Dam, and the extension of legal protection to the illicit gaming the syndicates already oversaw, the unheralded desert basin of Las Vegas suddenly began to sound like an appealing destination.

Guy McAfee was the first crook to set up shop. A commander in the Los Angeles Police Department's vice squad, McAfee had been forced to resign his post after a mayoral investigation found that he was operating a number of underground card games and that his wife was a Hollywood madam. McAfee was perhaps the first man to start over in Vegas, though he would hardly be the last. In 1939, he purchased the Pair O' Dice Club on Highway 91, the road that led into the city from the south. Admiring the expanse of desert around his little club, whose white stucco walls and Spanish-tile roof seemed to have been imported straight from California, McAfee optimistically labeled the road the Las Vegas Strip, a nod to the Sunset Strip in Los Angeles.

The Pair O' Dice was small-time compared to the operation overseen by Benjamin Siegel, who, by the early '40s, was already among the most feared organized crime figures in the country. As a teenager in New York City, he and a childhood friend, Meyer Lansky, had started their own gang rather than rise up the ranks of an established set. Siegel was said to have killed thirty people to ensure the ascendence of his crew through the 1920s, when its influence spread to the rest of the East Coast, and then Florida and Chicago, where the Bugs and Lansky Mob hooked up with the Capone family to become a national crime syndicate.

Siegel had matinee idol cheekbones and piercing blue eyes. He was unnervingly calm, too—until, without provocation, he flew into a rage. All of which is to say: he fit right in among the starlets and messianic producers of early Hollywood, where Meyer sent him after Prohibition was repealed in 1933. Siegel seized control of the gambling boats that anchored out beyond the LAPD's reach in the waters off Santa Monica and Venice Beach, then set up his own bookmaking operation, which churned through $500,000 every day. As the West Coast representative of a criminal organization concentrated east of the Mississippi, Siegel took it upon himself to get cut into the legal gaming business that was starting to gather momentum out in the Mojave Desert.

Initially, Siegel charged an underling, a fellow New Yorker named Moe Sedway, with ensuring that a wire service owned by their syndicate became the default choice in the legal casinos of Las Vegas. Sedway was so successful that Siegel was soon making $25,000 a month, just off fees generated by that branch of the wire. The legal casinos minted so much cash that Siegel saw that—for once—a sanctioned operation would outperform the shadow economy.

Siegel appealed to other influential members of his syndicate for financing. They agreed to float him several million dollars for construction. But even that wasn't enough for the Flamingo, which Siegel promised would be "the goddamn biggest, fanciest gaming casino and hotel you bastards ever seen in your whole lives." Not only would there be gambling, he said, but also a nightclub, restaurant, and luxurious pool, all spread across a thirty-acre parcel on the newly minted Strip.

As construction costs rose and his backers balked, Siegel found a new partner: Walter Bimson, the Phoenix banker *Ari-*

zona Highways would later canonize as one of the state's founding fathers. Bimson was the rare banker who had made out well during the Great Depression, capitalizing on the disinclination of other banks to make small loans to desperate clients. He became an early adopter of allowing installment payments on small loans, including automotive financing. He called his style of banking "poor man's capital," and was fond of saying, "I'd rather make ten loans for $1,500 than one for $15,000." While that democratic approach generated plenty of good publicity, the more lucrative line for Bimson's Valley National Bank was packaging all the interest that accrued from the small loans into a more traditional form of capital, the massive, high-upside loans it extended to land speculators like Del Webb, who spent the 1930s snapping up as much land in Arizona as he could.

Bugsy Siegel recognized Bimson as the rare banker who was likely to ignore conventional prohibitions against lending to clients embedded in the shadow economy. Working all possible angles, Siegel first hired Del Webb's construction company to work on the casino. When he inevitably ran short of cash, he appealed to Webb's benefactor Bimson, who agreed to extend the mobster a loan of $600,000, with another $600,000 kicked in from a financier in Utah, Walter Cosgriff. That Siegel had never run a straight business before in his life mattered less to Bimson and Cosgriff than the hitman's assurance that he was building a winner.

When the Flamingo opened the day after Christmas in 1946, it didn't quite live up to the hype: Siegel's inflexible plans for the debut meant that many of the rooms were not completed when cars began pulling up in front of the Flamingo's main entrance. Above the hotel's modernist wall of glass and the surrounding gardens stood a tall pillar, decorated with white neon spelling out the hotel's name—all of it positioned on the right side of the road three miles south of Las Vegas proper, making it the first and easiest stop after the long drive from Los Angeles: an oasis of leisure after 150 miles of the blacktop that tore through the Mojave's monotonous middle.

Bugsy Siegel wore a tailcoat with a carnation in the buttonhole on opening night, gamely ushering his guests to an amphitheater where the crooner Jimmy Durante was to perform. His jovial mood didn't last, given that far fewer VIPs had shown up than he anticipated. Apparently, William Randolph Hearst, still nurturing a grudge that began when both men were competing for the attention of a woman in Hollywood, had told celebrities to stay away lest he turn his newspaper empire against them. Siegel unleashed his fury after the show, when a genial tourist approached him. "Bugsy," the man said, "this is fabulous. God, what a beautiful place." Siegel—who loathed the nickname that nearly everyone knew him by—snapped back, "Who are you? Have we been introduced?" As the man stammered out an apology, Bugsy got even angrier. "How come you get so intimate?" he cried. As the man retreated with his wife and children, Bugsy kept screaming, "Get the hell out of here, you bum, and don't come back!"

Siegel couldn't get his head around the golden rule of Benny Binion, another of Vegas's early gaming pioneers, who said that the heart of the casino business was an effort to "make little people feel big." After its first two weeks, the Flamingo was down $300,000 and had to be bailed out by Bimson, who fully covered the early losses.

Through the spring, Del Webb made daily visits to Siegel's office in the Flamingo, taking a portion of the receipts to cover the hotel's construction debt. On one of those occasions, Siegel mentioned all the people he had personally killed, before laughing and assuring the businessman, "There's no chance that you'll get killed. We only kill each other." Siegel's associates in the syndicate were clearly impatient to start seeing a return on their investment, since they'd collectively loaned him far more to build the Flamingo than Valley National. In June 1947, Siegel's prophecy came true: he was found dead in Beverly Hills, killed by a bullet through one of his brilliant blue eyes.

Twenty minutes after Siegel's assassination, his old henchman Moe Sedway went to the Flamingo to meet up with another mobster named Gus Greenbaum, a Chicagoan who'd spent the last decade in Phoenix. The men assumed control of the Flamingo with a handshake. Greenbaum was appointed by the syndicate to run daily operations, which he swiftly turned into a moneymaker.

After the first year under Greenbaum, the Flamingo reported a profit of $4 million—not including the skim taken by its gangland creditors. The full-service gambling destination that Siegel had envisioned suddenly looked less like a flight of fancy than a model. For Vegas and, with time, the rest of America.

In short order, Las Vegas became a premier gambling destination. Eight million visitors spent $200 million in the city in 1954. Despite that windfall, success was far from guaranteed. The next year, not one but three resorts opened on the Strip—the Riviera, the Dunes, and the Royal Nevada—as well as the Moulin Rouge, the city's first integrated casino, which was located on the Westside. After only a few months, the Moulin Rouge had already closed and the debt-laden Royal Nevada needed emergency financing to keep the doors open. It, too, would go under in 1958. Despite all the evidence to the contrary, Wilbur Clark, the front man of the Desert Inn, enthused, "This town can absorb one new big casino-hotel a year for the next one hundred years . . . We're nowhere near the saturation point."

This sort of misplaced optimism pervaded Las Vegas when the Kim Sisters arrived in 1959. Luckily for them—because every Las Vegas story involves a run of luck—a new resort opened next door to the staggering Royal Nevada that year, and it badly needed help filling the thousand-odd rooms that gave it a claim to the title of largest hotel in the world. The Stardust was the brainchild of a San Francisco rumrunner named Tony Cornero who had used a savvy stock maneuver to raise $6 million for the project despite having only $10,000 to his name. Cornero's sleight of hand only came to light after he suffered a heart attack one morning while shooting dice at the Desert Inn. Afterward, a former associate of Al Capone known as Jake the Barber swooped in with $12 million to complete the casino.

The Thunderbird declined to compete with the Chicago muscle behind the Stardust, so after the Kims' original four-week contract was up they moved a block south and began performing there six nights a week. While the women's voices were no better or worse than any other lounge act's, the Kim Sisters were savants when it came to picking up new instruments. Not only could they memorize the words to any song by ear (none of them, at that point, spoke much English), they could do the same with drums, saxophones, and violas. In August, *Newsweek* brought the show's signature bait-and-switch to a national audience, describing how the performance began with "three prim Korean girls, robed in brightly oriental attire . . . playing a delicate Korean love ballad" before the trio unsnapped "their costumes to reveal slinky cocktail sheaths," and launched into rock 'n' roll numbers like "Fever" and "Charlie Brown." Around the same time, Ed Sullivan was in town scouting talent for his television review and was so impressed by the Kim Sisters that he put them on the air in September, where they reprised "Sincerely."

The Kim Sisters' first appearance on *The Ed Sullivan Show* in 1960 launched them into genuine stardom. Over the next twelve months, the group played the Waldorf Astoria in New York, bowled with Wayne Newton at Lake Tahoe, and attended the opening of Hugh Hefner's first Playboy Club in Chicago; when they realized that there were a number of nude women at the party, the Kims hid in a bathroom until their manager said the coast was clear.

Though they were growing in fame, Sue, Ai-Ja, and Mia (as Min-Ja decided to be called) still lived together at the unglamorous Robinson Apartments, a dingy complex right across the street from the Sahara. At the time, only around 120,000 people lived in Clark County, and Nevada had only shaken its longtime title as the least populous state in the union thanks to the admission of Alaska. Las Vegas proper was arrayed along a six-mile stretch of Charleston Boulevard; once you left the city limits going south, the handful of casinos that were beginning to fill in the Strip were flanked on both sides by the desert basin that remained as austere as it had for centuries.

Despite all the distractions the trappings of show business presented, Sue Kim remained focused on the task her mother had assigned her: to provide for herself and her family, even if that meant only taking one week off of performing over the course of an entire year. Ai-Ja, on the other hand, complained to the press that the group's performance schedule meant that "Our time not our own. No time to think!" and made a point of sleeping through rehearsals, even though it angered their manager.

Ai-Ja's chronic fatigue wasn't helped by her tendency to hang around the Stardust late into the night, making drinks for Frank Sinatra and chatting up Roger Moore.

Sue's demeanor was comparatively aloof. She was unimpressed by the other celebrities she encountered: when Elvis Presley came on to her after a performance in Tahoe, she brushed him off, remarking, "He was no different from anyone else." Her one indulgence was fashion. What money Sue didn't send home to her mother and siblings she spent on fur coats and jewelry. Sue would later tell her biographer, Sarah Gerdes, that the trio had no trepidations about the glamorous lifestyle they had developed almost overnight. "We were going to keep going forever."

By 1966, the infinite horizon the Kim Sisters imagined for themselves was starting to reach its outer limits. Ai-Ja befriended a member of the Korean Kittens, a dance troupe that had taken up residence in the Thunderbird's supper club after the Kims' departure. The two women stayed out late, drinking and gambling until they ran out of cash. Ai-Ja's decision to marry a drummer who had played for the Kim Sisters only led to more instability. Within two months of the wedding, Sue looked out the window of her new house on Desert Inn Road and saw Ai-Ja staggering down the street in a bathrobe, blood coursing from an eye that had been popped out of its socket.

Sue had suspected that Ai-Ja's husband, Frank Pastori, was abusive after noticing sores on her body, but didn't know how to intervene, even after he nearly blinded her. After a short break, the group returned to performing at the Stardust, and the cycle of violence continued: Ai-Ja would stay out late gambling and then be beaten by Pastori when she got home. Meanwhile, Mia and her husband abruptly left for Los Angeles, prompting Sue to take a look at the group's finances, which had long been handled by management. An accountant informed Sue that money was still being sent back to her mother in Korea, though she had been dead for several years. Not only that, but the commission paid to their booking agent, Tom Ball, who had also recently passed away, was still being diverted. "So I'm broke?" she asked the accountant, who shrugged his shoulders and said: "You're working week-to-week."

Ai-Ja finally divorced Pastori in 1972. The next year, the revamped Kim Sisters—with another of Sue and Ai-Ja's actual sisters, Jane, swapped in for Mia—moved to the Las Vegas Hilton, where they joined a roster of five lounge acts playing to an older crowd disinterested in the dance clubs that were coming into vogue. The Kim Sisters bounced from the Hilton to the Holiday Casino (eventually renamed Harrah's) and remained there for over a decade. No longer sought out by journalists and Hollywood actors, the musicians became just another outdated performance troupe coasting on nostalgia before finally retiring in the early '90s. Meanwhile, the Dae Han Sisters, another group of multi-instrumentalists with the surname Kim who had started out performing at Army bases in Korea, slid into the Hilton gig. "It is a very comfortable living," the bandleader, Yong, told an interviewer. "Maybe too comfortable."

The short shelf life of Vegas acts parallels that of the casinos themselves. Every few decades, a new impresario arrives in town who revolutionizes the gaming industry, inspires imitators, and then is surpassed and marginalized. So it goes across America, where constant cultural churn makes a fool of anyone who tries to overstay their welcome. The alternative is to root oneself in community, shunning the limelight in favor of a humbler but more sustainable life.

Not long before Christmas in 2022, every table at the Gritz Café in West Las Vegas was full of families and friends treating themselves to Saturday brunch. My server's name was Diamond; she wore glasses with pearls hanging off each corner and dropped off a platter of cornbread-crusted catfish, waffle on the side. One wall of the restaurant was decorated with two portraits, the larger an oil painting of a woman named Willa Mae Chaney, the recently deceased matriarch whose legacy included opening a community daycare, being elected to the Nevada Board of Education, and helping her daughter, Trina Jiles, found the restaurant. Next to her hung a screen print of Malcom X.

I was lingering on the Westside to get a better sense for the working-class Las Vegans who had made the city's rise possible.

Their history was on full display at Historic Westside Legacy Park, which was separated from the retrofitted bungalow that housed Gritz by the local FBI office, its perimeter defended by a serrated white fence and a pair of security gates. Beyond that complex, vacant lots flanked Mount Mariah Drive, over which a few golden orbs were visible, lofted above the park by tall, black columns. In the far distance, thin clouds had been caught in the clefts of the ridge of striated rock that forms the Sheep Range.

The path into Legacy Park ran along the edge of a Starbucks parking lot and was framed by a series of panels that reproduced historic photographs of the Westside: kids playing baseball and racing soapboxes, Muhammad Ali visiting Highland Elementary School in 1965, the facade of the Moulin Rouge. From there, the site opened out into a rectangle of tile and xeriscaping with a play structure to the right and a shaded area on the left, where a statue of Barack Obama reclined on a concrete berm. The geometric plaza he looked out on was studded with two-story-tall pillars, each holding a massive golden sphere.

Together, these columns constitute the installation *Living Black Pillars*, by Chase R. McCurdy. Though his family's history in Las Vegas goes back to the 1930s, when his grandfather, Herman Moody, first arrived in the city and eventually became the first Black officer in the Metropolitan Police Department's history, McCurdy followed a meandering road home. He left for college in Tucson, at the U of A, then lived in Los Angeles and Paris as he drifted from the corporate track into an artistic practice that encompasses photography, sculpture, painting, and poetry. For an exhibition of his work at UNLV's Marjorie Barrick Museum of Art, McCurdy cautioned visitors, "Please do not come (to my work) seeking a diversion or distraction from life. We are not at a point in history where that is an option." Instead, he wrote,

> *Let us merely focus on here*
> *if only for a moment*
> *for here is the seed*
> *the soil*
> *the root*

which gives bloom
to after.

Living Black Pillars suggests a sense of the "after." Reflecting sun in all directions as they hover in the clear sky, the golden orbs convey an electric possibility. At the same time, the foot of each pillar is surrounded by a series of plaques describing the lives of notable residents of the Westside, with plenty of space left over for new biographies to be added in the future. Because the columns are all individuated by a different colored accent, each feels like a vision of personal attainment that is grounded in the accomplishments of those who came before. All of it adds up to a neighborhood-wide project to uplift a population that has always had to fight for recognition from the city's power brokers.

Dozens of local Black luminaries are celebrated at Legacy Park. There's Donald Maurice Clark, the NAACP leader who campaigned for integrating the Strip; Joe Neal, the first African American elected to the Nevada State Senate, where his advocacy for the neighborhood earned him the nickname "Westside Slugger"; and the pioneering dancer Anna Bailey, who first came to Las Vegas to perform at the short-lived Moulin Rouge. "Everything was springing up then, more hotels and more businesses and everything," she recalled of the Westside in the late 1950s. "There was so much hope." Herman Moody is included among the honorees, as well as the labor leader Hattie Canty. Next to her plaque is one for Ruby Duncan. In the engraved portrait, Duncan is smiling, her sprightly eyes surrounded by a corona of curls.

Ruby Duncan was among the thousands of Black women who moved to Las Vegas from the Mississippi Delta during and after World War II. She came from a family of sharecroppers outside of Tallulah and began working the cotton fields there at just eight years old, only permitted to attend school in the four months between picking and planting season. After mechanized cotton pickers were introduced, even that meager livelihood vanished. Duncan's extended family joined the hundreds of thousands of other Black southerners who were heading west during the so-called Second Great Migration. Duncan followed an aunt to

Nevada in 1952, eventually securing an apartment in a public housing development called the Cadillac Arms, on the other side of the railroad tracks from downtown Vegas.

The Cadillac Arms was one of the few apartment complexes on the Westside at the time; the rest of the neighborhood was made up of trailers, canvas tents, and wooden sheds, all laid out neatly along dirt paths. Few homes had swamp coolers or municipal plumbing, and almost none were large enough for a bathtub. In the summer, young men would cart around metal basins, charging their neighbors twenty-five cents for a dip. The neighborhood didn't get paved roads or a real sewer system until 1955, basic infrastructure that was funded through a deal with the federal government to build I-15 along the old railroad right-of-way, but which only served to further isolate the Westside from the rest of the city.

Despite the atrocious conditions they were expected to live in once they arrived, migrants continued to flock to Las Vegas from the Deep South, and the city's Black population shot from 178 to more than eleven thousand in the space of two decades. Though Strip casinos depended on Black cooks, housekeepers, and bellmen, they were not allowed on the premises after their shift ended. That meant nightlife for Black Las Vegans revolved around Jackson Avenue on the Westside, which hosted clubs like the Town Tavern, El Morocco, and the Brown Derby. When the Moulin Rouge opened on nearby Bonanza Road in 1955, Black entertainers who were disallowed from mingling with the guests on the Strip flocked there. Sammy Davis Jr., Harry Belafonte, and Dinah Washington all made after-hours appearances in the Moulin Rouge's lounge, while part-owner Joe Louis worked the front door. The Moulin Rouge got so hot that white performers started showing up as well, and soon Frank Sinatra, Rosemary Clooney, and Bob Hope were all whiling away their evenings on the Westside, chatting up locals at the bar under a mural of Parisian can-can dancers. That June, the cover of *Life* featured the Moulin Rouge's own showgirls in colorful fringe dresses and yellow-feathered headpieces.

The Westside's renaissance was short-lived. In October, the Moulin Rouge closed when its principal owners, already behind on their debt payments, failed to make payroll. Unlike with the mobbed-up casinos in town, no white bankers were interested in floating the Moulin Rouge until it became profitable. With their favored hangout shuttered, frustrated Black entertainers used their fame to forcibly integrate the Strip. Harry Belafonte sat down at a blackjack table at the Sands and wouldn't budge until he was dealt in. Around the same time, Anna Bailey led a group of Black women to the same casino to see Frank Sinatra perform. A bouncer was about to throw them out before Ol' Blue Eyes himself intervened, leading the women to his private table. In 1960, a pair of doctors from the Westside, James McMillan and Charles West, threatened to stage a march on the Strip if ordinary Black visitors weren't afforded the same treatment as the celebrities. The casino barons, wary of how a protracted standoff could interrupt their cash flow, capitulated. Unfortunately for the clubs on Jackson, their clientele suddenly had many more options for a night out, and the commercial district fell into steep decline.

By then, Ruby Duncan was working as a maid at the Flamingo and had married an airman from Nellis Air Force Base named Roy. The family included seven children, as Duncan, like many Black women of the era, was repeatedly denied a prescription for contraception. Only after her seventh child was born was Duncan able to convince a doctor to tie her tubes, and only then because she grabbed the headboard of her hospital bed and refused to budge until the procedure was performed.

Duncan's husband became physically abusive shortly after the couple moved into a standalone house on the Westside, but every time she tried to walk out with the kids he sweet-talked her into staying. "He tried to be a good father," Duncan told the historian Annelise Orleck. "I was interested in keeping my children happy and clothed and fed. And he was a good provider." The kids themselves had their own opinion of the situation. When Roy attacked Duncan during a particularly intense argument, the older kids revolted. One grabbed a mop, another a broom, and a third picked up a kitchen knife from the counter. Outnumbered, Roy bolted out onto the street and never looked back.

With Roy gone for good, Duncan sought out a better-paying

gig in the kitchen of the Sahara, but was forced to give it up after she slipped in cooking oil and injured her back. Hoping to find a job that didn't involve standing, Duncan signed up for a sewing class at the local welfare office that aimed to prepare beneficiaries for work as a seamstress. There, she met a host of other West-side women with similar backgrounds—practically all of them were from the South, had numerous children and no partner, and struggled to get by on the pittance provided by Nevada's welfare system, which was led by George Miller, who, less interested in helping the needy than in ferreting out cheats, likened welfare mothers to "vultures."

Over the buzz of sewing machines, the women shared their welfare horror stories, taking regular breaks to fan themselves as the temperature rose in the stuffy office. Dentistry and optometry were considered "cosmetic" and not covered, which meant the degenerative eye condition of one of the women's daughters went years without treatment. Officials would often raid their homes to see if a man was present—if so, he'd be designated as a "substitute father," regardless of his relationship to the children, and all benefits would be revoked.

Nevada was hardly the only state where welfare mothers were treated with hostility. As governor of California, Ronald Reagan called welfare a system that incentivized single motherhood, vowing, "From now on, the able-bodied will work for their keep," even as he and the rest of the ascendant conservative movement ignored the dependence of their rural constituents and corporate allies on different forms of governmental largesse. An Albuquerque mother named María Avila once described the reality of the lifestyle politicians like Reagan portrayed as a free ride:

There I was, seventeen years old with a kid and no job, living on welfare. The welfare gave me a hundred and twenty-eight dollars a month for me and Tasha. My rent was a hundred and ten. The utilities brought that to a hundred and fifteen. Ten dollars went for food stamps, and that left three dollars for whatever else we needed. I didn't have enough money for soap or to take the bus to the food stamp office, nothing!

Meanwhile, Congress was authorizing direct payments to farmers struggling to make a profit in the free market and slashing corporate tax rates. Remembering the feeling of connecting with other mothers who had been trapped by a program designed to permanently sequester them to the margins of society, Ruby Duncan told Orleck, "It felt good . . . All those things that had been locked up inside us for so long finally started to come out."

Just before Christmas in 1970, George Miller's office issued a report claiming that more than half of the welfare recipients in Nevada were "cheaters" who were receiving more money than they were due, "with twenty-two percent completely ineligible." Though the state provided no documentation to back up these claims, 4,500 families saw their benefits cut and another 3,500 were jettisoned from the program entirely over the next month. By then, the group of mothers who first met in sewing class had organized. As the newly minted Clark County Welfare Rights Organization, they began contacting other families who had lost their benefits from Miller's random audits. Duncan's phone rang constantly as women whose financial lifeline had been severed reached out for help. The local Legal Services office filed suit against the state, but in the meantime, the welfare rights activists set out to marshal public opinion on their behalf.

George Wiley, the founder of the National Welfare Rights Organization, flew to Las Vegas from Washington, D.C., to help coordinate a mass demonstration. After Duncan picked him up at the airport, they drove the length of the Strip, with Duncan narrating, "This is the main vein of Nevada. This is *the* pocketbook." Together, the two activists planned an unprecedented march, one that would make good on the threat James McMillan and Charles West had made to force the integration of the casinos a decade earlier. The night before the action was scheduled, organizers assembled at UNLV. Many shared their concerns about the sort of violence they could expect from casino security should they try to disrupt business. Duncan spoke up, utterly transformed from the woman who had once needed her children to defend her. "If we march down the Strip, someone could get hit in the head with a brick," she conceded. But, Duncan continued, fear was no reason to shy away from the fight. "When you're poor, and you have to send your children to school without food, or food money or decent clothes, that's a violence we live with every day. To get hit with a brick one day really wouldn't mean anything."

The next morning was March 6, a Saturday that dawned sunny and clear, though the air retained a winter chill through the early afternoon. Hundreds of welfare families assembled in front of Circus Circus, where they met sympathizers who had journeyed to Las Vegas from all over the country to march in solidarity. Duncan and Wiley led the column, a crowd of several thousand people that included Cesar Chavez, Ralph Abernathy, and a group of Ojibwe welfare recipients, who had traveled from North Dakota and wore traditional buckskin clothing. After two miles, the march veered toward Caesars Palace. As the protestors entered, the panicked staff of a boutique scrambled to conceal a display of fur coats while players at the table games lunged to secure their chips.

All gambling at Caesars came to a halt; across the street, the Flamingo followed suit. Abernathy paused near a statue of the casino's titular Roman emperor and appealed to him with a playful, impromptu sermon. "I can't obey the laws of this land when they conflict with the laws of God. Caesar, what are we going to do?" Duncan laughed uproariously at the performance, while other women lounged in red banquettes to watch their children splash around in the fountains. For the first time, the women who had spent years working on the Strip to facilitate someone else's leisure were able to enjoy a resort for themselves.

Though the march that followed a week later was far smaller, it was arguably even more disruptive. Finding the doors of the Sands locked, the 250 activists who had descended on the hotel linked arms in the middle of Las Vegas Boulevard, blocking traffic for hours. Mary Wesley, one of the original mothers involved in the founding of the Clark County Welfare Rights Organization, was working on the Strip that day but quit on the spot so she could go sit on the asphalt in solidarity.

Days after the second demonstration, a federal court ruled that George Miller's bid to shred the welfare system in Nevada had been illegal and ordered the full restoration of benefits. In a subsequent investigation, the U.S. Department of Health, Education, and Welfare found that Miller had fabricated the numbers in the report that justified culling the rolls, and determined that his employees were responsible for many of the erroneous claims that had been uncovered. To celebrate their victory, the welfare mothers threw a party at the defunct Moulin Rouge, complete with catering. Sammy Davis Jr. picked up the bill.

The Moulin Rouge never did reopen, even as it remained a community gathering place until 2003, when it was gutted by arsonists. A few years later, what remained of the building burned to the ground. Today the lot is nothing but sand and parched grass sandwiched between a boarded-up store and an industrial facility. Jackson Avenue isn't much better, lined with more empty lots than storefronts, most of which are vacant. I found the street itself closed when I drove over from Legacy Park, as the city had just begun a six-month makeover that would include repaving, widening sidewalks, and adding streetlights. I parked on a side street and walked over the concrete foundation of a building that had been torn down, only a cinder-block outline and some degraded tile indicating where a structure had previously stood.

Across Jackson, a Baptist church shared a wall with the Soul Brothers Motorcycle Club; on the next block, the old Town Tavern was long shuttered. Given all the empty windows looking out on a street made of sand, the scene felt like an eerie echo of the early '50s. Still, the wall of the building next to the concrete platform I was standing on had been painted blue for a half-finished mural that read "Change is coming—Historic Westside." The "is" hadn't yet been filled in, though butterfly wings were painted under the words, with a gap between them to create a future photo op with colorful handprints of community members extending on either side.

The dilapidated state of Jackson Avenue was a painful reminder that, however inspiring, the welfare rights movement had only managed to temporarily delay the bipartisan effort to pare back the social safety net, culminating in 1996, when Bill Clinton proclaimed the end of "welfare as we know it." The welfare money the federal government provides to each state has been frozen since then; in rapidly growing states like Nevada and Arizona, that means once-meager benefits are now practically useless. Nevada received just $63 per child in 2019, compared to $409 in California. A Nevadan named Danielle Frolander told ProPublica that her benefits were down to $50 a month. "It's kind of silly, these amounts. It goes into my gas tank to get to work, and that's about all."

However downtrodden, the Westside has hardly been defeated. When the city broke ground on the repaving of Jackson, the neighborhood's city councilor, Cedric Crear, declared, "We are going to continue to advance our community, whatever that takes. If it's five years, ten years, thirty years, fifty years, we are going to be here."

A few days before visiting the Westside, I met up for dinner with my friend Sreshtha at the Oyster Bar. It's an eighteen-seat restaurant that's open 24/7 inside Palace Station, a resort just off the Strip that originated what is now a chain of sixteen casinos that stretch across southern Nevada and explicitly cater to locals rather than tourists. Given its limited seating and outsize reputation, there's almost always a line. After Sreshtha greeted me wearing a pink, fur-collared jacket and a Guido-appropriate gold chain around their neck, the two of us waited for over an hour to sit down. A guy wearing expensive leisurewear had enough time to leave the line to get a beer from a different bar, return, drink it, and go get another one before he sat down.

Meanwhile, Sreshtha and I chatted about their writing students at UNLV and watched the action on the floor, where every table was crowned by a flat-screen TV playing a Patriots-Cardinals game. One poker table hosted a grizzled old man, a silent lady in a surgical mask, and another woman with long false nails who clapped after every hand; next door, a middle-aged guy in a sweatshirt played blackjack. When the time came for him to fork over more cash, the dealer counted it out by spreading the twenties and hundreds on the green felt between them before converting the bills into chips. A cheerful passerby leaned over and asked, "How's she doing? She lucky?" The player grumbled something under his breath, and the man addressing him retreated. Two tables down, a jittery dude was playing three hands at once, chain-smoking, and monologuing to the expressionless dealer. All of a sudden, there was a loud cheer a few tables back and a man sprung up from his chair and elaborately fist-pumped while his friends celebrated. After that, the usual, quiet hum of people losing money imposed itself again. Now at the front of the line, Sreshtha pointed out a handwritten sign that read "No oysters today."

Once we sat down at the oyster-less Oyster Bar, the couple next to us asked the chef in the open kitchen what had happened. He was an Italian American who looked like he was on his thirty-second restaurant in about as many years. With a conspiratorial smile, he replied, "Trust me, you don't want to eat them." The chef was performing as much as cooking, dropping jokes while heaping butter and broth into a silver pot that looked like the graphic at the end of a slot machine rainbow.

By the time Sreshtha and I had paid and were getting ready to leave, it was past eleven o'clock, but the line for the Oyster Bar was longer than ever. Unfazed, the chef kept his pots roiling and the blackjack dealer maintained her steady rhythm. After doling out the cards, she'd check her hand by nudging one corner off the green felt in an action so minimal as to be imperceivable. Then, once the haggard gambler in the sweatshirt called, she'd use a single, fluid motion to flip her hand face up and reach out, pulling ever more chips back from player to house.

If you have a job in Clark County, there's a one in four chance it's in hospitality. The key feature of that industry—which encompasses everything from gambling to restaurants to hotels to strip clubs—is its accessibility: you don't need a college degree to cook, or clean, or entertain, you just need diligence and drive. That means Las Vegas is every bit as working-class as Philadelphia or Cleveland—it just happens to be a lot more fun.

Over the course of the 1990s, the city's largest union, Culinary Workers Local 226, doubled in size, to more than forty thousand members, with nine thousand of them joining in just 1997 and 1998. That the union had managed to make such strides in tandem with the booming casino industry led the then president of the AFL-CIO, John Sweeney, to declare in a speech that "just as surely as New York set the standards for the past hundred years,

Las Vegas will be setting them for the next hundred years." The
city presaged a novel era of labor organizing, wherein baristas,
line cooks, and knowledge workers all over the country forged
a new consciousness of themselves as a class within an economy
whose orientation had long since transitioned from industry to
service.

The occasion for Sweeney's visit to Nevada was a rally in
support of the workers at the Frontier Hotel and Casino, who
maintained a strike from 1991 to 1998. Among the longest labor
actions in American history, the Frontier strike only ended after
the hotel was sold and the new owner agreed to raise pay and
cover back wages. Midway through the work stoppage, Local
226's president, Hattie Canty, proudly told the *The Washington
Post*, "Las Vegas, whatever else anybody says about it, has become
a city where maids can own their homes and raise their families."
Canty had grown up in Alabama and worked as a housekeeper at
the Thunderbird in the early 1970s. After her husband died, she
became the sole provider for eight children. She found a new job
at the newly opened Maxim Hotel as a uniform attendant and
soon realized how much she relied on her union benefits. In an
oral history, she recalled a doctor finding a cancerous growth in
one of her son's ears. "If I had not belonged to the union and if
I had not had that Culinary insurance," Canty said, "I wouldn't
have been able to get that operation."

Canty became president of the local in 1990, the first Black
woman—and maid—to hold that position. Beyond overseeing the
successful strike against the Frontier, she is most remembered for
starting the Culinary Training Center, which students can attend
for free thanks to an agreement with hotel managers, who recognize the value of a pipeline of well-prepared employees. "We
teach people how to prepare food. We teach people how to serve
food," Canty said in 2007. "And most of the time, ninety percent of our girls or ladies that go through that Culinary Training
Center get jobs." The school remains such a vital component of
the city's labor infrastructure that when the Culinary Workers
endorsed Hillary Clinton for president in 2016, the event was
held on its premises.

When I spoke to Ted Pappageorge, who was elected to lead

Culinary Workers Local 226 in 2022, he told me, "When the rubber hits the road, you've got to make things happen and you've got to do it in the street." He equated casinos to "service factories," and said that, as in the steel mills of old, "workers here have the ability to do anything as long as we're united."

Pappageorge is a lifelong Las Vegan. His mother was a server, his father a bartender who worked his way up to bar manager. Pappageorge started out as a busboy at the Sands when he was sixteen and ended up bartending at Binion's Horseshoe downtown after doing a few semesters at UNLV, though he never got a diploma. His first contact with the union came in 1990, when the workers at the Horseshoe went on strike. "We were in trouble. Our union had lost six casinos in 1984, when there was a city-wide strike. In 1989 the industry came back, essentially to wipe out the union. I really didn't think we were going to win when we went on strike, I just knew we had to fight." After Jesse Jackson and Rich Trumka, the president of the United Mine Workers, came to Las Vegas to support the walkout, Local 226 shifted away from thinking solely in terms of the day-in, day-out of their campaign. "We realized we weren't just fighting for Las Vegas or just for our union. It was a mentality that workers could stand up and win." After nine months, the Horseshoe's management finally agreed to a new contract.

Pappageorge soon went full-time with the Culinary Workers, where he helped to manage the six-year-long Frontier strike. Together, those two labor actions inaugurated a broad-based approach to union organizing that has continued into the twenty-first century, helping workers in every industry—from the arts to agriculture to college athletics—feel empowered to organize even when the legal system has been intentionally stacked to thwart them. Indeed, one survey from 2017 found that roughly half of nonunion workers nationwide would vote to organize their workplace if given the opportunity.

Though the Culinary Workers are far stronger in the 2020s than they were in the 1980s, Pappageorge is clear-eyed about the challenges of the contemporary labor environment. Beyond reaching the ten thousand nonunion workers who staff independent restaurants on the Strip (some of them improbably nestled inside union casinos), he's also hoping to organize the staff of the city's massive convention center and its nascent sports arenas. "Those are low-wage, shitty jobs. They're part-time, where people go from event center to event center to try to get enough work to equal one full-time job. Those should be union jobs. People should be able to get health care and own a home." I was speaking to Pappageorge just a few months after three hundred workers at the convention center had voted to authorize a strike, and he suggested that they were only a week from walking out. "Look, we're not going to back down. People are fed up."

Despite Las Vegas's proud history of labor organizing, visitors still struggle to understand the city as anything beyond an overgrown theme park. The hundreds of thousands of workers who make it possible for casinos to rake in upward of $12 billion every year are effectively invisible within the neon-lit fantasy they commute to every day from North Las Vegas, Spring Valley, and the Westside. Some of those laborers are thriving, some suffering, some crooked, some straight. Each of them is driven by the desire to make a life for themselves in a place whose very existence encourages the belief that, through force of will alone, the desert can accommodate any desire at all.

Few observers have endorsed that dynamic quite as proudly as the art critic Dave Hickey. In his view, the appeal of Las Vegas is no more complicated than the fact that it is "a town where outsiders can still get work, three shifts a day, around the clock, seven days a week—and, when not at work, may walk unmolested down the sidewalk in their choice of apparel." The city's greatest detractors were Hickey's colleagues at UNLV, "lost souls" who "do *not* arise from their slumber thanking their lucky stars to have escaped Mom and Dad and fucking Ithaca.

"Most of all," he continued, "I suspect that my unhappy colleagues are appalled by the fact that Vegas presents them with a flat-line social hierarchy . . . being a *professore* in this environment doesn't feel nearly as special as it might in Cambridge or Bloomington." Hickey's diagnosis explains my own restless relationship with Cambridge, where I have lived longer than anywhere else outside of New Mexico yet still feel out of place. For all its quotidian charms—great bookstores and bike lanes, lovely parks and outstanding Indian food—the city is still dominated by the byzantine mores of academia, a system meant to rebrand run-of-the-mill social climbing as a meritocracy.

No matter how discomfiting I might find the dollar-worshipping ethos of Las Vegas, at least it's honest. This is a city where ostentation is an unambiguous virtue. Nobody dresses down: the point of making money is to look like you have money. If the Kim Sisters spent too much on fur coats in their early years, it was only because their milieu demanded it. How could they command the next big booking fee without appearing as if they were already stars? Notoriety is so analogous to wealth in Las Vegas that it would be nonsensical to pursue one without the other. Naturally, that means the most famous residents have always been the ones who get to deposit their customers' lost wages at the end of the night, and no one has ever become quite as synonymous with Sin City as Steve Wynn.

Wynn's first visit to Las Vegas was as a child in 1952, when his father, who ran a couple of bingo parlors in upstate New York, was in the midst of a failed bid to open one up at the Silver Slipper. "It was a cowboy place with a strange kind of exotic energy," Steve Wynn remembered in an interview with Charlie Rose. "I think people came to Las Vegas as much to see the characters as they did to gamble and to see the shows. The place itself was a spectacle. And a very, very overwhelming one, too."

Wynn returned in 1967. Having inherited the bingo parlors, he bought 3 percent of the Frontier Hotel for $75,000; within months, another of the Frontier's owners, Jack Shapiro, was busted meeting with Anthony Zerilli, the son of a Detroit mob boss. Luckily—there's that word again—the idiosyncratic billionaire Howard Hughes had just bought the Sands and was looking to gobble up more real estate on the Strip, so he was glad to take over the Frontier despite the legal heat.

Around the same time, Wynn became acquainted with a banker from Utah. Though he grew up in the Church of Jesus Christ of Latter-Day Saints, E. Parry Thomas played a central role in rebranding gambling as a reputable business, easing the transition of Las Vegas from a city run by far-off criminal syndicates to one whose fortunes traded on the New York Stock Exchange. Another Mormon banker, Walter Cosgriff, had chipped in some equity when his idol, Walter Bimson, bankrolled Bugsy Siegel's Flamingo Hotel in the '40s, and Parry Thomas would finance sin to an even greater degree in the years to come, asking, "How are you going to do banking in a state by ignoring its main industry?"

Thomas had a hand in every mobbed-up casino of the '60s, including Moe Dalitz's Desert Inn, the Patriarca family's Dunes, and Meyer Lansky's Thunderbird. In the space of two years, his firm, the Bank of Las Vegas, saw its assets compound from $250,000 to $16 million. Backing for the loans that generated those returns came from Cosgriff's Continental Bank and Deseret National, which had been founded by Brigham Young himself. Every Monday, a plane was chartered to carry the millions of dollars Las Vegas had generated for the church over the previous week back to Salt Lake City.

Thomas became a locally beloved Jack Mormon, though his wife and children remained devout. Even-keeled and reliable, Thomas could talk business with anyone, regardless of their reputation, but he took a special liking to Steve Wynn, telling *Fortune* that the young man "was quick, honest, and had an exceptional IQ." In other words, however unproven, Wynn cut the profile of a traditional real estate developer much better than the banker's usual clients.

Shortly after the two men met, Thomas financed Wynn's bid to buy some land back from Howard Hughes and then flip it to Kirk Kerkorian (then the owner of the International Hotel) for twice what he'd paid. Wynn netted $700,000 from the transaction, which Thomas advised him to use to buy a stake in the Golden Nugget downtown. Within two years, Wynn had edged out the other owners, and, at just thirty-one years old, he became the youngest person to ever run a casino in the state of Nevada. Of Parry Thomas, Wynn would later reflect, "I think I can trace every one of the wonderful opportunities [I] had to the advice of this man." So great was Wynn's affection for Thomas that he called him "Dad"; for his part, Thomas referred to Wynn as "my fifth son."

After taking over the Golden Nugget, Wynn's theatrical persona became the perfect mirror to the city he sought to dominate. In 1983, Wynn cut a TV ad for the casino that opened on him in a natty suit and his signature black helmet of hair. He stood in one of the hotel's suites, which was decorated with a crystal chandelier and a white piano, and pompously expounded on the amenities. When he strolled over to Frank Sinatra—in a dashing pink sport coat—and introduced himself as the owner of the hotel, Sinatra peeled a dollar bill off his bankroll and pressed it into Wynn's hand, telling him, "You see I get enough towels." Cut to a close shot of Wynn, mugging for the camera like the beleaguered hero of a sitcom, his gravelly voice pitched high: "Towels?!"

Wynn's image as a kingpin who didn't take himself too seriously grew from there. Ahead of the opening of Treasure Island in 1993, the writer Mark Seal described approaching the developer's inner sanctum through a succession of brightly painted rooms. The office itself was "absolutely electrifying," with magenta chairs and a carpet "streaked with rivulets of primary colors." "And leaning back in a white leather chair with armrests big as hotel pillows," Seal wrote, "his white tennis shoes kicked up on his desk, wearing blue jeans and a black Donald Duck T-shirt, is Steve Wynn." He drove a chocolate-colored Maybach, wore a Patek Philippe, and spoke to his beloved German shepherds exclusively *in* German, but on the Wynn Las Vegas's opening night in 2005, he was still happy to work the crowd, posing for a picture whenever a guest called out for his attention. Unlike Bugsy Siegel, Wynn had taken the advice of Benny Binion to heart: he knew how to make the little guys feel big.

The personal touch mattered because Wynn had cast himself as Las Vegas's leading citizen, never having forgotten his childhood realization that the energy of the city flowed from the residents who chose to make a spectacle of themselves. When his first Strip resort, the Mirage, opened in 1989, it included a dolphin pool where Wynn could often be found swimming with his pets in a wetsuit. Then, to distinguish the Bellagio as a high roller's retreat from the riffraff that gravitated to the new roller coasters and cartoonish architecture of the Strip's "family friendly" era,

Wynn went on an art-buying binge, dropping $285 million in the space of two years. As Wynn explained to the journalist Christina Binkley, "It's a terrible mistake to underestimate the taste of the public."

Through Wynn's rise, his greatest rival was the Los Angeles–based Kirk Kerkorian, whose investments in Las Vegas dated back to the 1960s, when he had gobbled up undeveloped property on the Strip that eventually turned into Caesars Palace and the Las Vegas Hilton. By the late '90s, Kerkorian owned the MGM Grand and New York–New York but kept a low profile, overseeing his casinos from an unmarked office building in Beverly Hills. While Wynn was all sizzle, Kerkorian didn't use credit cards or own a sports car, preferring to motor around in a Jeep Cherokee.

Kerkorian had been keeping close tabs on the mounting debts of Wynn's company as it spent $2 billion to build both the Bellagio and Beau Rivage, an ill-advised resort in Mississippi. Once the Bellagio had proved its staying power, Kerkorian made his move, confident that Wynn's shareholders would leap at a takeover offer. He was right. Despite the enormous impact the Bellagio had on both the casino business and the international image of Las Vegas, after only two years Wynn had lost control of his crown jewel, along with the Mirage, Treasure Island, and the Golden Nugget, in a deal worth $6.4 billion.

Within five years, Wynn was back, with the titular casino he built on the site of the old Desert Inn—a landmark of Old Las Vegas which the man he called Dad had helped finance. From his new foothold on the Strip, Wynn's resurgent empire spread first to Macau and then to Boston before *The Wall Street Journal* reported allegations from dozens of employees that he had engaged in what the paper called "a pattern of sexual misconduct." The story included accounts of Wynn routinely exposing himself to female employees, forcing a massage therapist to masturbate him, and raping his manicurist.

When Wynn resigned, he was forced to sell his 12 percent stake in the company, a transaction that netted him over $2 billion. His childhood friend D. Boone Wayson, whose father had worked with Wynn's at one of those bingo parlors in upstate New York, succeeded him as chairman of the company and issued a statement on behalf of its board that credited Wynn for "transforming Las Vegas into the entertainment destination it is today." Though the statement claimed that Wynn Resorts was committed to "being an inclusive and supportive employer," that was the closest it came to acknowledging the reasons for Wynn's departure.

Sin City has seen many catastrophic falls after a meteoric rise, but Steve Wynn's story is a necessary reminder that not every Las Vegan hits the ground in the same way. Though Sue Kim made out okay, Ai-Ja died tragically, her cheeky charisma utterly drained by years of abuse, addiction, and gambling. That dynamic—the women suffer and the guy gets paid—is familiar across American life. The only difference in Vegas is how its bald-facedness illustrates the undercurrent of violence that animates so much of the city's culture and history. That so many Americans still look askance at Las Vegas might have something to do with how untroubled the place seems to be about its least savory aspects. As Hunter S. Thompson put it, "In a closed society where everybody's guilty, the only crime is getting caught."

The profound racket of Fremont Street was audible from four blocks away, ricocheting between the banks of empty office buildings and over the vacant sidewalks of downtown Vegas. Turning the corner from Fourth Street onto Fremont and its electric canopy, the sound became outright deafening. There were high-end buskers everywhere: a woman singing R&B, a steel drum group, a contortionist putting himself inside a knee-high safe to the tune of '90s electronica. Louder than each performer's individual soundtrack were the snarling yelps of Lil Jon, which permeated Fremont Street thanks to the massive speaker banks of the permanent stage situated in the plaza between two casinos, the Four Queens and the D. Wearing a leather thong and a green fur coat, the "Dancing DJ" dialed down the music to ask for people in the crowd from Oregon and Florida to cheer, since a college football bowl game was scheduled to take place between their teams that weekend. The hollering was more persuasive when she asked who might instead be visiting from California, an excuse to queue up

Katy Perry's "California Gurls" and get back to romping around the stage.

Overhead, the Fremont Street Experience's arced canopy extended for four blocks and was seeded with 49 million LED lights, churning like a high-def, psychedelic screensaver. As the prevailing color of the screen shifted from red to green to yellow to blue, the complexion of the crowd underneath it changed accordingly, bolstering the sense that, as with all spectacles, the onlookers were an integral part of the show. I stopped for a moment to look around, and an older couple wearing orange Oregon State fleeces wandered by, both of them fixated on a preteen rapping into a single-speaker microphone next to a two-story-tall Christmas tree. I got moving again once the stench of weed had gotten too overpowering and walked for a minute behind a burly dude in a T-shirt that read "Pipefitters Local 537—Boston, MA," a backward Patriots cap on his head. The Dancing DJ flipped the soundtrack to Shakira, and suddenly the canopy was all tropical fish and coral.

With a dramatic whoosh and a few screams of delight, five tourists flew by overhead on a zip line as the people on the ground held their phones up to record the daredevils shooting from one end of the Experience to the other. Beyond the main stage, two chiseled men wearing only cowboy hats and bolo ties above their enormous belt buckles and blue jeans smiled warmly at every passing woman. In front of a gift shop, a man whose face mask was striped in the colors of the Argentine flag stood near a glowing space heater with a massive python wrapped around his shoulders—he didn't seem to be soliciting tips, only watching the crowd. A dude in a blazer bought a cigar from a stand in the middle of the action and immediately lit up, puffing the smoke in my face as he strode away.

Two women were having their IDs checked by the bouncer of the Circa, the city's most recently opened casino and the first to be built from the ground up in downtown Vegas since 1980. One of them was bopping so aggressively to "Hips Don't Lie" that the presentation of her driver's license became an excuse to shimmy her chest only a few inches away from the security guard's. When it was my turn, I told him that, while I'd be happy to hand over my ID, I wasn't exactly prepared to dance. He chuckled and waved me inside.

The Mega Bar extended down the length of the casino's main gaming floor, the entire wall behind it tiled with big-screen TVs that jittered with the constant movement of hockey and basketball. Even those LCDs were dwarfed by the screens in the Circa's sports book, a literal amphitheater occupying the back end of the building, with eleven rows of stadium-style seating and fourteen luxury booths facing a three-story-tall, 78-million-pixel display. Booking one of the recliners for a big match required a minimum spend on food and booze of a couple hundred dollars; during football season, renting a booth on Sunday cost $2,500, plus a mandatory $500 tip.

That night, there were just some desultory regular-season games—the Lakers hosting the Nuggets in L.A., the St. Louis Blues playing in Calgary. Still, a fair number of seats were taken, with a dozen guys loitering at the semicircular bar above the bowl to check the over-under odds. Many more guests were gathered around the table games, which were organized in circles, with a pit boss standing behind the dealers in the center of each cluster. Next to him, a woman danced on top of a box wearing a flapper-esque outfit consisting of a low-cut fringed top and a miniskirt in matching blue. Nearly all the woman dealers wore the same get-up. Every twenty minutes, the dancers would step off their boxes and trade places with a dealer, each woman efficiently segueing from spinning a roulette wheel or collecting chips to casually gyrating with a blank expression on her face.

The soundtrack had been perfectly calibrated to the nostalgic inclinations of the sharply dressed Millennials who were crowding around the tables and peering curiously at slot machines. Being one of them, I circulated around the casino for far longer than I'd intended, nodding along as the music shifted from Walk the Moon to Rihanna to "Island in the Sun" to "Timber," by Pitbull and Kesha—effectively the same playlist I would've heard at a party in college. Both gaming floors were oriented around a three-story well where a twenty-five-foot-tall neon icon of old Fremont Street had been restored and mounted: Vegas Vickie, a busty cowgirl in white chaps and gloves whose boot, as a nearby placard explained, had "kicked playfully over crowds for more than three decades."

Vegas Vickie originated as an advertisement for Bob Stupak's Glitter Gulch, an early entry in the long-running campaign of the visionary showman to make an impact on the city, which finally found its culmination with the Stratosphere, the thousand-foot-tall mashup of the Seattle Space Needle and CN Tower that still dominates the Strip's skyline. When I spoke with the Circa's CEO, Derek Stevens, he credited Stupak as one of a handful of gambling impresarios who informed what he hopes to accomplish downtown, along with Steve Wynn, Jay Sarno—the mastermind who slipped the apostrophe out of Caesars Palace since he wanted all his guests to feel like monarchs—and Ben Siegel (even seventy years after his assassination, nobody with juice in Vegas dares use the moniker "Bugsy"). "I was always a bit of a Las Vegas historian," Stevens said of the years when he was first getting involved in the casino business, a period that coincided with Steve Wynn's bid to turn Las Vegas into what Stevens called "the preeminent hospitality city in the country" with the opening of the Mirage. "And then, boom, Steve Wynn took it to the Bellagio, which really changed Las Vegas into a broader-reaching community that addressed arts and international travel."

Stevens and his brother, Greg, first bought into the Rio and IGT, a slot-machine maker, in the '90s, two components of a national investment strategy that grew out of their success running an auto parts company in their home state of Michigan. Though conventions and trade shows had first drawn Stevens to Vegas, the city's sports gambling culture is what really captured his imagination. "I'm a guy that likes math and loves sports," he told me. "I'd never want to miss a Super Bowl, I'd never want to miss a Wild Card Weekend, a World Series game. A boxing match. I *love* boxing."

In 2006, the brothers bought a stake in the Golden Gate, one of downtown Vegas's oldest casinos, and then took it over entirely four years later. Meanwhile, Stevens gradually transitioned from spending most of his time in Detroit to fully living in Nevada. "I kind of moved to Vegas one Delta trip at a time." The Stevens brothers financed a massive renovation of the Golden Gate and started populating its casino floor with Dancing Dealers, then did the same down the street at the Fitzgerald, which they purchased and renamed the D in 2012.

The branding of the D was a nod to both downtown and Detroit; Stevens funded an ad campaign for the hotel across the Midwest and opened an American Coney Island chili dog stand inside. Big and boisterous, Stevens became a regular presence at the D's bar after the hotel reopened. He bought a white Shelby GT350 Mustang, then held a seven-month-long promotion called "Win Derek's Car." Slot players could enter into monthly drawings, and at the end of the process thirty of them assembled at the Fremont Street Experience and were each given boxes, only one of which held the sports car's key. At the event, Stevens wore a blazer with its sleeves brocaded in gold. A few years later, he made a sharp bet on Michigan State to win the college basketball championship, placing $20,000 at fifty-to-one odds. Duke prevailed that year, preventing Stevens from cashing a million bucks, but that didn't matter: after Michigan State made the Final Four, ESPN and *The New York Times* wrote stories about the bet, introducing Derek Stevens to a national audience for the first time. "We didn't have enough money to hire a PR firm or a spokesman or a pretty face to put on a camera. I'm the cheapest guy around, so let's go do it."

With the Circa, Stevens put all the most salient features of the first two casinos—the Dancing Dealers, the D's Long Bar, the Golden Gate's sense of history—to work in a high-gloss resort oriented around the entertainment landscape of the twenty-first century. "In the '90s, there was must-see TV. People would change their schedule to be able to watch *The Cosby Show* or *Cheers. Friends.*" That, of course, changed with the advent of streaming. "But nobody's going to record a sporting event and watch it tomorrow. No one talks about the Super Bowl on Tuesday. Sports is a different kind of entertainment that's always going to be watched live—and it could be watched together." Circa began to take shape around the idea of "bringing people together in the world's largest sports book."

Since the casino opened in 2020, Stevens has been a regular presence in the amphitheater, hosting a digital talk show for Circa's football-betting Survivor Contest in a salmon-colored suit and even jumping into the dogpile of a group of fellow Detroiters after the Lions beat the Packers in the last game of the season. Surely, he was the only casino owner to ever get quite that excited about owing his customers a million dollars. It's impossible to imagine the CEO of Caesars Entertainment, Anthony Carano, or Bill Hornbuckle, the head honcho at MGM Resorts, working the floor of either of the conglomerates' numerous properties, let alone doing so dressed up as a cheerleader or in a sequined sport coat—just two of the outfits Stevens has been photographed wearing. In an era where management of most of the gaming industry has been concentrated among a few boardrooms staffed by aspiring Kerkorians, Stevens is an outsize character in the Wynn mold, equal parts sports-crazy oaf and accessible bigshot.

"In the end," the philosopher Bruce Bégout wrote of Las Vegas, "it is all so much sensational overkill that you wind up with a feeling of utter stupefaction." Elsewhere in his book about the city he complained, "Even if you make every effort to merge into the overexcited atmosphere of the place, you still can't stop yourself from experiencing a feeling of deep unease . . . The sadness probably derives from a contrast between the quantity of energy expended, in all innocence, in order to join in the party, and the signal poverty of the outcome."

That night at the Circa, I wondered if the dude needed to lighten up. For all the celebrity residencies, Cirque shows, and sports books, the greatest attraction in Vegas has always been the city itself, one of the few places in America where you can just hang out for hours on end. The Strip and Fremont Street collectively function as an infinite mall, a place that is so overwhelmingly profitable in its totality that it doesn't actually matter if any individual visitor spends any money. You're never far from a comfortable chair, a spacious public bathroom, a TV broadcasting a football game, a musician performing. Everywhere you turn, something interesting is happening, strangers are mingling, and a sense of possibility suffuses the air. Though bosses like Stevens enjoyed most of the profits, the workers were getting by all right too—Las Vegas has become one of the few cities in America where service work is a sustainable career, one that can provide a home, health insurance, and a comfortable life.

In 1990, the authors of "The Sunbelt Syndrome" asked, "What does freedom mean outside of a social context? Is it freedom to experience a new lifestyle rather than to create a new social structure or a new community? This kind of freedom suspends values and leads to a sense of anomie." What they refused to understand is that the lifestyle they saw emerging was not antithetical to community, it was a different version of it. After all, there is no high culture or low culture in the Southwest—just culture. Those who ignore the region or refuse to acknowledge its status as quintessentially American do so out of a misplaced belief that social position or intellect can somehow elevate you above the rude reality of capitalism. Las Vegas feels particularly threatening to the *New Yorker* set because it refuses to indulge their pretensions: as Dave Hickey pointed out, there is no social currency here, just money. It's in this way that the desert lays bare the hardscrabble nature of America, exposing the nation as a place where power has always been its own pursuit.

My friend Laura and I spent an afternoon hiking near Zia Pueblo, climbing a crest of white rock known as the Dragon's Back. The spot is popular with mountain bikers—so many had recently ridden the ridge that their tire tracks had left divots in the dust, revealing the brilliant gypsum just under the surface. Off to the east, the formation fell away steeply into a ruddy bowl, only a few scrubby bushes clinging to the rust-colored earth. The other slope was a little gentler, and we eventually found a comfortable rock where we could sit and watch a thunderstorm approach Cabezon Peak, the flat-topped laccolith fourteen miles to the west. Dark virga tendrils hung from the underside of the cloud, a promise of approaching rain, while bolts of lightning jittered down into the earth. We stayed until the smell of damp soil reached us, then booked it back to the car, stopping only long enough to let a tarantula cross the path, each of its furry brown legs probing the crumbles of gypsum for purchase as it hustled toward shelter.

On the drive back to Albuquerque, we stayed on the west side of the Rio Grande, taking the state highway that passes through Rio Rancho. The area had been known as the Alameda Land Grant in colonial times, bestowed on Francisco Montes Vigil by Governor José Chacón

Medina Salazar y Villaseñor in 1710 to recognize Vigil's participation in "all of the efforts to pacify the Indians." The grant comprised the one hundred thousand acres of land opposite the Sandia Pueblo; this was far more than Vigil actually needed, so he flipped his title to the land to Juan Gonzales two years later, making a tidy profit of two hundred pesos. The Gonzales family retained most of the Alameda Land Grant through the Mexican-American War, after which one of the heirs, Antonio Lerma, petitioned the United States for formal recognition of the family's ownership. What followed was four decades of surveying reports and legal wrangling, leading up to 1911, when the land grant was "partitioned" by a court, with the 450 residents splitting ten thousand acres and the rest being put up for auction and eventually acquired by Anglo ranchers. In 1961, the real estate developer AMREP bought the fifty-five-thousand-acre Koontz Ranch and began marketing it as the Rio Rancho Estates. Most of that land was subdivided into seventy-seven thousand lots, which AMREP sold for over $200 million.

By the time Laura and I were in high school, Rio Rancho was an established suburb of Albuquerque and one of the fastest-growing communities in the country, surging past Santa Fe in the rankings of New Mexico's largest cities. The boom ended in 2008, and the impact of the housing market's collapse on Rio Rancho was obvious and visceral. Four miles from the developed areas of town nearer the river, an arena had been marooned—it had briefly been home to the New Mexico Scorpions, a minor-league hockey team that folded after the market crashed—along with a municipal office, a call center, and a hospital.

Driving down Paseo del Volcan, the cluster of buildings that forms what Rio Rancho insists on calling City Center seemed like a mirage. It was twilight when we reached the cluster of buildings. We did a quick lap around their freshly paved but empty parking lots, mostly just to verify the reality of the place before heading east, toward the Rio Grande. Navigating off the map on my phone, I instructed Laura to turn right. The normal-looking cross street ended abruptly: suddenly, we found ourselves in an arroyo. As Laura laughed and reversed back up the dampened gravel of the incline, I tapped my phone to switch from the standard map to a satellite image. The gridded-out streets all vanished, along with the names of subdivisions and master-planned developments. In their place, there was only open space.

COMBUSTION

The Southwest is a place where almost everyone slips their moorings and just drifts. The cities and towns are ugly, the populace footloose, the crime frequent, the marriages disasters, the plans pathetic gestures, the air electric with promise.

—CHARLES BOWDEN

For most of his childhood, Jimmy Santiago Baca lived in Estancia, a ranching village on the Llano Estacado, the 150 miles of arid plain that bridges New Mexico and Texas. Whiling away endless hours in his grandparents' weathered homestead, the seven-year-old Baca would sit and watch thunderheads build over the parched flatlands. At night, he'd lie awake listening to the unearthly moaning of the wind against the house's rickety walls.

Baca was born in Santa Fe in 1952, but landed up in Estancia after his father abandoned the family and his mother, Cecelia Padilla, began dating an Anglo named Richard. She bleached her hair, instructed her children to use only English when Richard visited, and began pointing out "white-skinned, blue-eyed children" to them as an example to be followed. Whenever Baca or his siblings asked their mother for chili or tortillas, Richard would snap, "It's time you started eating American food."

The breaking point came when Richard permitted Cecelia to bring her children along to a visit with his parents. On the drive over, he turned to Cecelia in the passenger's seat and informed her that "it would be best" if she told them she was Anglo and that the kids belonged to a friend, that she was merely babysitting them. In his memoir, Baca recalled watching his "mother bite her bottom lip as she stared straight ahead. I expected her to say something back to him, but instead she said to us, 'You better be on your best behavior.'"

The very next day, Cecelia and Richard took the children to their grandparents' shack in Estancia. Cecelia looked at her children, assuring them, "I'll be back in two weeks. I'm going on vacation and I'll be back in two weeks to pick you guys up." Two weeks passed—nothing. After another two weeks, an enormous box arrived, containing all their clothes and toys from the house in Santa Fe.

In his poem "Silver Water Tower," Baca captured the sense of pride he developed about living in Estancia under the care of his grandparents, where he felt wanted,

> where the link of my true blood was unbroken,
> and thrived solidly in a little silver hair'd woman,
> with ancient customs, y pura sangre nuestra;

Such heady thoughts of familial continuity were dashed by the exigencies of the present. The children's stay in Estancia only lasted a few years. In 1959, their grandmother died and there was no other choice but for him and his brother Mieyo to be sent away to Albuquerque to live at a Catholic orphanage. Baca's aunt delivered them to the somber, brick building, just across the street from the massive Albuquerque Indian School, a complex that harbored children from the Southwest's pueblos and other tribal settlements.

Both institutions had been designed for the express purpose of deracinating the kids in their care. At St. Anthony's, boys slept under a leaky roof that soaked their bedsheets every time it rained and were beaten by nuns for any perceived misbehavior—including speaking Spanish. There was little educational instruction. Instead, kids were put to work: tending to the farm on the grounds around the facility, doing laundry, or cleaning the chapel. It was a place where impoverished children, most of them Mexican American, were groomed for a subservient role in society. To those, like Baca, who objected, the nuns made it clear that "I didn't belong, I didn't fit in, I was a deviant."

St. Anthony's had opened in 1913, three decades after the Albuquerque Indian School. Like the rest of the federally administered Indian boarding schools, that institution was modeled on the Carlisle Indian School in Pennsylvania, whose founder, Richard Henry Pratt, had infamously proclaimed that his mission was to "kill the Indian . . . and save the man." In 1885, the Albuquerque Indian School's first superintendent, R. W. D. Bryan, wrote, "The ultimate object of the Indian schools is, as I understand, not so much the improvement of individuals as the gradual uplifting of the race." In practice, that meant preparing students to assume menial jobs for white employers. Most of the classes were forms of "industrial education": boys were taught stonecutting and carpentry, girls to sew and cook. A man from Taos Pueblo named Jerry Suazo attended the Indian schools in both Albuquerque and Santa Fe in the 1930s and recalled being forced to bathe, cut his hair, and wear clothing provided by the government, a boot camp–like environment that even included roll call.

Many children fell ill or were injured at the Albuquerque Indian School; those who did not survive were buried on the premises in unmarked graves. In 1973, a city construction crew uncovered human remains while installing a sprinkler system at nearby 4-H Park. At the time, a former caretaker named Ed Tsyitee told the *Albuquerque Journal* that as many as thirty students had been buried during his three decades at the school. The residents of St. Anthony's were likewise subject to neglect, if not outright victimization. In 2020, a man alleged having been sexually abused by the chaplain for seven years, beginning when he was six years old. Soon after, more former residents went public with their own stories of trauma. In a statement, the first man's attorney told reporters, "We believe there are many other victims of sexual abuse who were trapped at the orphanage as children with nowhere to go, who are still silently suffering out there."

In "Silver Water Tower," Baca wrote of his speaker's arrival in Albuquerque: "I rebelled, no con mi mente, / pero con mi Corazon." Something about St. Anthony's superseded his ability to reason, and instead made him feel, deep in his heart, that he simply did not belong in this new world of strict rules and dutiful studies. He tried to run away. Every time, he was caught by police, returned to the orphanage, and beaten by the nuns. Each subsequent escape attempt elicited more severe punishment: eventually, three sisters took him to a private room, instructed him to drop his pants, and took turns spanking him with wooden paddles. Baca looked over his shoulder at one of the nuns, who was sweating from the exertion. She snarled, "We're going to break you, kid." Contumacy taking root in his chest, Baca thought to himself, "This is nothing. I'll be gone tomorrow."

That night, Baca waited until the lights had gone out in the dormitory. He crept downstairs, eased open an old door, and began picking his way through a field of alfalfa, doing his best not to rustle the leaves. Once he finally found the fence, he reached up to grab a bar and heaved himself over and into the sharp drop-off of an arroyo that abutted the property. Ignoring the pain of the fresh welts on his backside, Baca took off down the ditch, following it downhill toward the Rio Grande.

This time, Baca managed to evade being returned to the orphanage—instead, he ended up at a juvenile detention center. Once he was released from *that* internment, Baca found himself living in "a gardener's shack, more or less like the one I'd grown up in." Less than a decade after their arrival in the Duke City, "My brother and I were alone in the world," Baca wrote in his memoir. "I was fifteen and he sixteen, and we were accountable to no one." While Mieyo started working at the Desert Sands Motel on Central, Albuquerque's main drag, Baca had trouble replicating even that modicum of stability. "Security guards and managers followed me in store aisles; Anglo housewives walking toward me clutched their purses as I passed. I felt socially censured whenever I was in public, prohibited from entering certain neighborhoods or restaurants, mistrusted by government officials, treated as a flunky by school teachers, profiled by counselors as a troublemaker, taunted by police, and disdained by judges, all because I had a Spanish accent and my skin was brown."

By the late 1960s, the congenital division between New and Old Albuquerque had become a gulf. Though the two cities were now one, Mexican Americans and Indians in Albuquerque were straining under the new order. As the Chicana poet and scholar Gloria Anzaldúa wrote: "This land was Mexican once, / was Indian always." In seeking to resist the conditions imposed upon them by the ruling class, many Chicanos and Indians took inspiration from the Black organizers who had fought and defeated Jim Crow, as well as rural activists elsewhere in New Mexico who were forcibly reclaiming the land that had been seized by the U.S. government after the Treaty of Guadalupe Hidalgo.

A century of Anglo control had marginalized the peoples who had called New Mexico home for far longer than that. Now, the dam that held back decades of defeats and disappointments was beginning to crack under the pressure of resentment. What followed was an awakening that precipitated the Southwest's gradual reshaping into a region that took pride in its multicultural reality instead of grasping at the white supremacist fantasies of its founding fathers.

When a census taker visited the home of a man named Ventura Chávez on the Westside of Albuquerque in 2002, he asked, "Are you Hispanic, sir?" Chávez's grandson, the scholar Simón Ventura Trujillo, remembered the look of befuddlement that settled over the administrator's face when he received the response: "Hispanic? No, we are Indianos. Some people have called us Indo-Hispano. We are mixed, half Native American and half Iberian, mostly Jewish"—that last flourish a reference to the prevalence of so-called Crypto Jews among the early colonists, those who practiced their faith in secret during the Spanish Inquisition. Chávez continued:

> We are not "Spanish" because "Spain" was not a sovereign country when Columbus came here. And we are not *only* Iberian. My great-great-grandmother was half Apache and half Cherokee. She became a slave to the Spaniards and was named Dolores Lucero. We are mixed.

The census taker checked "other" and went on with his day. On another occasion, Trujillo remembered his grandfather cutting his finger while chopping onions. "There went the Iberian part of me," he joked.

Racial and ethnic politics in New Mexico are much more complex than an opposition between brown and white, or even white and everyone else. Nor is the state a place where the contentiousness of finely grained differentiation is shunned in favor of some sense of the universal. Clearly, Trujillo was not the sort of person who would claim a line of Spanish ancestry that extended back to the conquistadors in hopes of persuading Anglos of his whiteness. Instead, he leaned into the confusion and contradictions of mestizaje heritage. This sort of embrace was fundamental to the Chicano movement, which emerged in the 1960s not to replace old racial categories with a new one, but rather to highlight the common experiences of the oppressed peoples of the Southwest so they might rise as one against the Anglo and Hispano ruling class.

Albuquerque first became central to El Movimiento in 1963, when Reies López Tijerina came to the city with eight hundred other activists for a convention that marked the 115th anniversary of the Treaty of Guadalupe Hidalgo. Tijerina was a radical Tejano pastor who had become fixated on the history of the Spanish land grants during a visit to northern New Mexico in the mid-'50s. He had learned of how, in addition to ending the Mexican-American War, Guadalupe Hidalgo guaranteed the right of former Mexican subjects living in the areas ceded to the United States to both citizenship and the land that had been allocated to them by the old land grants, which covered nearly all of New Mexico as well as parts of southern Colorado and Arizona.

Over the next few years, Tijerina educated himself about the Santa Fe Ring, a coterie of lawyers who gravitated to New Mexico after the Civil War and managed to turn over most of the land grants to Anglo investors through a combination of esoteric legal maneuvering and outright fraud—many of the original Spanish documents were destroyed or simply disappeared. As land grants like Tierra Amarilla, which surrounds Chama, shrank from 580,000 acres to ten thousand, a handful of men claimed title to vast swaths of New Mexico, including Thomas B. Catron, who managed to acquire title to more than 2 million acres.

To fight for the restoration of the land grants to their right-ful heirs, Tijerina founded the Alianza Federal de Mercedes. Though the Alianza was mostly a rural campaign, Albuquerque's media outlets became its megaphone, with Tijerina broadcast-ing a daily, fifteen-minute speech on Spanish-language radio and appearing every week on local television.

At the same time the Chicano land reclamation campaign was gaining momentum in town, Burque became a central organiz-ing point for the Red Power movement. The University of New Mexico's Kiva Club was one of the first Indian student organi-zations in the country, and, in 1968, the National Indian Youth Council moved its headquarters from Berkeley to Albuquerque so that it could better coordinate with the tribes of the Southwest. The next year, Gerald Wilkinson, a Cherokee Catawba lawyer, took over the NIYC and, over the next two decades, facilitated the establishment of dozens of satellite offices across the country in cities and college towns with large Indian populations, from Phoenix to Stillwater, Oklahoma. The NIYC's network of grass-roots organizers made it a vital player in a number of local tribal actions, including when it joined with the Coalition for Navajo Liberation to stop Anglo corporations from building coal gasifi-cation plants on the Colorado Plateau.

The proximity of Chicano radicals and Indian activists in Albuquerque led to many fertile debates. When Indian activists pointed out that the Spanish land grants represented little more than a form of colonial seizure, Tijerina shifted the argument toward the need for continent-spanning opposition to the win-nowing of both Chicano and Indian territory by Anglo oppres-sors, saying that "as our land grants were taken away here in New Mexico, in Mexico they were taken away from the Indians and the new breed [of mestizos] by blue eyed Spaniards . . . in Argen-tina the same, in Brazil, in Chile, Peru." Whether or not that particular line of thinking proved persuasive, it was inarguable that both groups were pushing back against the same forces.

Powerful Anglos considered the radical ferment of 1960s and '70s Albuquerque as a bunch of hippie-adjacent nonsense, insisting that the city ought to conform to the American main-stream: suburban and segregated within a clear racial hierarchy.

The municipal model these boosters looked to for inspiration was Phoenix. Though about the same size as Albuquerque in 1950, the Arizona capital's growth far outpaced the rest of the region over the next decade. By the late '70s, Phoenix's manufacturing output was more than $2.6 billion, compared to $363 million in Albuquerque; as far as agriculture went, the shortfall was even more substantial. Phoenix generated twenty-seven dollars for every one dollar eked out of the Rio Grande. In response to the widening gap, Albuquerque formed a citizens' committee of local money men who sought to exert as firm a hand on municipal politics as Phoenix's chamber of commerce did. But their efforts to lure in new employers were hampered by Albuquerque's crime rate, then among the highest in the country. One local official leaned on stereotypes to explain the city's situation, telling the *Albuquerque Journal* that "much of the crime we are facing is being spawned in the minds of those who are idle and consequently, lack the means of obtaining the things they want and need."

This sort of attitude ignored the persistent discrimination minorities faced in Albuquerque, including those seeking to better their economic standing by getting a college education. The Chicana activist Patricia Luna recalled to an oral historian the "blatant racism" her darker-skinned classmates at UNM experienced. "Some were among the brighter students," she said. "They'd make A's on tests and get C's in courses. There used to be signs on lawns, 'Apartment for rent—no Mexicans or dogs.' During those years my sense of outrage kept growing and growing." Newer residents of Albuquerque were far more interested in carving out a comfortable life for themselves in the prefab Northeast Heights than they were in finding common cause with anyone living in a sunbaked trailer down by the river, allowing the predominantly Chicano and Indian communities in the more historic parts of town to fall into deeper economic degradation.

Whatever hopes Albuquerque's boosters might have had of replicating the trajectory of urban Arizona, they were often stymied by the democratic process. In 1966, voters successfully defeated a "slum clearance" effort that would have leveled much of the South Broadway neighborhood—the historic center of Albuquerque's Black community. Two years later, the city pro-

posed an even more ambitious plan, the Tijeras Urban Renewal
Project. Much like the Pueblo Center Project in Tucson, this
plan sought to demolish the homes and businesses on 182 acres
just north of downtown Albuquerque to make way for a new civic
plaza, including a $2 million Hall of Justice. Voters resoundingly
rejected that bond measure, too: unsurprisingly, half of the city's
population was not interested in displacing people who looked
just like them. In response, the city appealed to the federal government for the money, which was happy to oblige, releasing
most of the necessary funds in 1970. By the end of 1975, roughly
two hundred buildings just north of Central Avenue had been
replaced by fifteen new structures, including a convention center
and a new police headquarters.

This was the environment in which Jimmy Santiago Baca
came of age. Though he was often too preoccupied with the simple matter of survival to be more than peripherally involved with
politics, Baca deeply identified with El Movimiento. In an essay
written ahead of the five-hundredth anniversary of Columbus's
voyage to America, Baca reflected, "I have no sense of participation in the national identity of this country, no sense that I am
represented." His mother was descended from the Comanche-
ros, the mestizo settlers who made their living trading with the
Comanche at fairs in Pecos and Taos, even as the tribe was routinely pillaging settlements along the Rio Grande throughout the
eighteenth century. His father's parents, on the other hand, were
what Baca called Mexican "peasants." If an identity was available
to him in America, it was surely that of the Chicano, since, he
wrote,

> Chicano culture is young, vibrant, ever-reaching and
> risking, and each new step we take renews our sense of
> who we are. I am constantly discovering who I am, how
> my Spanish and Indian heritages speak to each other of
> anger, forgiveness, and love, creating a third voice, the
> Chicano voice.

As a teenager, Baca learned how determined the world was
to silence that voice. The activist Antonio Moreno once esti-

mated that a dozen Chicanos were killed in confrontations with
the police across New Mexico in the late 1960s, including Bobby
Garcia, who was found in an Albuquerque arroyo with a bullet in
the back of his head. Around the same time, the Black Berets—a
militant Chicano group modeled on the Black Panthers—learned
that its membership had been infiltrated by informers for the
Metro Squad, a secretive intelligence-gathering operation that
was a joint effort of the New Mexico State Police and several
local agencies, including the Albuquerque Police Department.
A federal civil rights investigation later concluded that Chicano
organizers were routinely subjected to secret wiretaps, warrant-
less searches, and other forms of aggressive surveillance. It also
detailed several police killings that suggested a lack of due process, including that of James Bradford, a Black man who was
killed by a lone police officer at Kirtland Air Force Base in April
1971.

Making matters worse, law enforcement had no qualms
about meeting peaceful protestors with violence. In 1970, UNM
students who had occupied a campus building to protest the killing of four of their peers at Kent State University in Ohio were
forced out by National Guardsmen wearing identity-obscuring
gas masks. When the students resisted, the soldiers affixed bayonets to their rifles, stabbing ten students and a cameraman from
a local TV station and leaving the concrete walkway outside the
Student Union smeared with blood.

The profound tension between activists and police came to a
head in June 1971, when a free concert at Albuquerque's Roosevelt Park, just south of the university, was broken up by police,
who arrested some participants for smoking weed. Outraged by
the apparent attempt to punish Chicanos just for socializing on a
nice summer evening, someone in the crowd yelled, "There's the
pigs. Let's get the pigs!"

What followed was an uprising the *Tribune* called "the wildest night in Albuquerque's history": police were far outnumbered
by the crowd of around a thousand concertgoers, who broke into
a police cruiser to liberate the guys who had been arrested. The
crowd then heaved itself against the cruiser, forcing one side off
the ground and then scattering as the car fell on its roof with

a massive crunch, broken glass skittering out in all directions. The crowd cheered, then pushed over two more police cars. As authorities retreated from the scene, a handful of more politically minded Chicanos gathered a group to march to the city hall.

The crowd passed under the overpass of I-25 and then across the bridge that rose over the old AT&SF railroad tracks, before turning north. It was the weekend, so there were few cars to interrupt the column's progress downtown, their voices echoing across the mid-rise office buildings as they chanted slogans like "¡Tierra o muerte!" and "¡Raza sí! ¡Guerra no!," references to the land grant struggle and the Vietnam War, where so many young Chicanos were being sent to fight for a government that treated their families as disposable.

Once the protestors crossed Central, they hit the Tijeras Urban Renewal Project site, where demolition had already begun on several buildings. Finally reaching the city hall, some members of the crowd chanted and sang, while others attempted to break into the locked building. A few managed to get in and set small fires inside; out on the sidewalk, others threw rocks through the upper windows. By this time, the police had regrouped and began firing tear gas into the crowd, which only succeeded in spreading them out across downtown. The protestors took over Central, with a few even pretending to be traffic cops, mawkishly directing cars through the bedlam. Dozens of businesses were looted, with several buildings going up in flames overnight.

The unrest continued through the next day. This time protestors converged on Yale Park, closer to the campus of UNM. As a crowd of around two hundred people marched past an Oldsmobile dealership on Central, a demonstrator threw a rock through its window. A few others delighted at the destruction and hefted their own stones, while a salesman pleaded with a policeman who was keeping his distance, "My god, do something!" Hundreds of National Guardsmen were deployed, but that didn't stop every storefront in one three-block section of Central from having its windows smashed, or another twelve buildings from being set on fire. By the end of the night, another 238 people had been arrested.

After the rioting subsided, the AP carried a photo of two

Guardsmen standing in front of a pile of stones and shattered timber—the remains of a school administration building. There was no rhyme or reason to the firestorm that gripped Albuquerque for two days, just a mass of humanity lashing out in frustration at authorities who expressed bewilderment at the senseless violence. In fact, many Anglos blamed "out-of-state" agitators for the riots, without offering anything in the way of evidence. Lieutenant Governor Roberto Mondragón was roundly criticized for telling protestors, "I know we have had police brutality. It is not alleged but factual." An editorial in the *Albuquerque Journal* sniffed that this statement "may have gained cheers from his listeners but we cannot recall a single instance of alleged police brutality."

A week after that *Journal* editorial, a nineteen-year-old Santa Fean was killed by a "warning shot in the back of the head" while his hands were handcuffed. Six months after that, two members of the Black Berets, the journalist Antonio Cordova and the anti-carceral activist Rito Canales, were murdered by the Metro Squad. An informant for the police later testified that he had invited Cordova and Canales to meet at a construction site on the Westside of Albuquerque. After they arrived, officers in hiding opened fire.

In his long narrative poem "Martín," Jimmy Santiago Baca summed up the opposing sides at Roosevelt Park this way:

> *When we burned*
> *a cop's car to the ground*
> *He clubbed a Chicana for talking back.*

Despite the enormous power of disruption implicit in a gathered crowd, the more intimate violence of the police prevailed, as it always does. Order was eventually restored—or at least, the version of it that allowed for the continuing operation of car dealerships and jewelers—even as the core complaints of disrespect and inequitable treatment that had led to the riots went unaddressed.

Given the seeming impossibility of ever being treated with dignity, some Mexican Americans resolved to do what they could to pass as white—including Baca's mother. Baca learned that, after dropping him and his siblings off in Estancia, she had eloped to San Francisco with Richard and begun calling herself Sheila. The couple eventually returned to New Mexico with two children of their own, settling in a white neighborhood in Albuquerque. Baca's sister managed to reconnect with their mother, and eventually persuaded the teenage Baca to pay her a visit. After he arrived, "Sheila" introduced Baca to her new children as a "friend," and then proceeded to ask him formal, bland questions in the kitchen.

"We remained standing," Baca wrote in his memoir. "She didn't talk about herself, and she didn't need to; the evidence of the good life was all around her, from the expensive leather and wood furniture, the new refrigerator stocked with food, the sparkling pool beyond the glass doors I could see leading to the patio." Baca responded to the rejection with a "three-day binge of whiskey and drugs." The speaker of "Martín" follows a similarly self-destructive path.

> *I lived in the streets,*
> *slept at friends' houses, spooned*
> *pozole and wiped up the last frijoles with tortilla*
> *from my plate. Each day*
> *my hands hurt for something to have.*

He presents his speaker

> *on a corner, beneath a smoking red traffic light,*
> *I live—*
> *blue beanie cap snug over my ears*
> *down to my brow,*
> *in wide bottomed jean pants trimmed with red braid*

Martín is

> *entangled in the rusty barbwire of society I do not understand,*
> *Mejicano blood in me spattering like runoff water*
> *from a roof canale.*

It would be decades before the blunt attempts by Anglos to sideline, subjugate, and suppress Chicano and Indian voices gave way to a less overt brand of predominance. In reality, the city's shifting demographics proved more powerful than the aspirations of its turn-of-the-century boosters. At the same time a tide of former GIs and midwesterners seeking sunshine were moving to the Northeast Heights, the largely Hispanic and Indigenous population that lived in the rural expanse of New Mexico was also gravitating to the city, where jobs were plentiful and housing was cheap. Even some of the old-timers from Plaza Vieja who had left for California during the war years were eventually drawn back, retiring to an Albuquerque that, despite its profound changes, still felt like home.

Once the census began tracking "Spanish origin" in its decennial surveys, it became clear that the American metropolis New Albuquerque's founders had envisioned had indeed emerged, but it was entirely unlike the place they imagined. In 1980, Albuquerque was found to be only 56 percent Anglo. By 2000, the city had become majority-minority.

Baca fled Albuquerque for California shortly after the reunion with his mother, but eventually returned home after a chaotic interregnum. He was picked up by police for a murder he had nothing to do with and arrested for a variety of minor offenses, all of it leading to a five-year confinement at the maximum-security prison in Florence, Arizona, after he was caught up in a drug raid at the border. At that point, Baca was an illiterate twenty-one-year-old, having never attended enough school to learn how to read or write.

Baca educated himself in prison. After a Good Samaritan from Phoenix randomly chose him as a pen pal, Baca carefully studied his letters before embarking on the arduous process of writing back. He composed lines and images in his head before slowly committing them to paper in his cell, using a pencil he sharpened with his own teeth. "A stream of ideas flowed through me, but they lost their strength as soon as I put them down," Baca wrote in his memoir. "I erased so often and so hard I made holes in the paper. After hours of plodding word by word to write a clear sentence, I would read it and it didn't even come close to what I'd meant to say. After a day of looking up words and writing, I'd be exhausted, as if I had run ten miles." Still, he kept at it. "I can't describe how words electrified me. I could smell and taste and see their images vividly. I found myself waking up at 4 A.M. to reread a word or copy a definition." Soon, he was writing poems. Baca's first chapbook was released in 1978, the year before his release.

"I had been scooped up and taken to prison as part of growing up as a Chicano," Baca said the first time we talked on the phone. "Not because I was guilty of anything, but because I was a Chicano, and this was the way the pipeline went." Indeed, Baca was one of the hundreds of thousands of Chicanos ensnared by the age of mass incarceration. By 2001, almost ten times as many Hispanics had served time in prison compared to the rate in 1974. Baca described his incarceration as "a ritual coming of age." Though his circumstances were "bleak" when he got out, he felt drawn back to Albuquerque, comforted to return to a place "where I saw people who looked like me and understood the beauty of our food, our customs, our language, our rituals, and our heritage."

Baca settled in the South Valley. Then, as now, the South Valley was an unincorporated extension of the city where active farms and wandering chickens provided a sense of continuity between the contemporary city and the agrarian province of Tiguex. When Baca arrived, the neighborhood's three-hundred-year-old adobe farmhouses and newer subdivisions had become home to many muralists and other Chicano artists. "There I could hear the music, and eat the food, and participate in the dances and the painting and the weaving and the pottery. There was this really lively culture that was connected to Mexico, and Mexico was connected to Central America, and it went all the way down to Chile. All these people coming and going, like this underground road where people from Chile were coming up to Albuquerque and people from Albuquerque were going down to Chile and Mexico, Oaxaca, and Chiapas."

Baca found a house in the South Valley, a "seriously wrecked fixer-upper" that some drug users had been crashing in. One day in 1983, when he was up on a ladder repairing the stucco of the house's ceiling, his mother pulled up outside in a Lincoln Continental. Decked out in a red dress and with a fistful of diamond rings, she confided to her son that she "had not spoken Spanish in so long she had almost forgotten it." Overwhelmed with guilt by how her second marriage had forced the rejection of both her original family and her heritage, she was thinking of filing for divorce, but was afraid to do so because, she told Baca, Richard had been saying "for a long time that he'll kill me if I leave him. But I can't lie anymore. I have to tell my children and his parents the truth. Only then can I start living my life."

Within a week, she had begun the process of confessing her origins to the entire family. "She was in her kitchen polishing her nails, preparing to go dancing," Baca wrote, "when Richard came into the kitchen and shot her in the face five times with a .45."

Like the divide between Old and New Albuquerque, the Chicano uprising of 1971 isn't widely discussed in New Mexico. Perhaps that's because the convulsions of the late '60s and '70s didn't settle the question of New Mexico's contested identity, let alone reorder the power dynamics that pervade American life.

I feel the persistence of that gap every time I'm home, driving on streets that feel little changed after almost two decades away. Even as portions of the Central corridor and the North Valley have contended with gentrification and the attendant spike in housing costs, much of the city remains exactly as it was when I was in high school. Same businesses, same potholes, same problems. The Albuquerque Police Department remains one of the most notorious local law enforcement agencies in the nation, while tens of thousands of nonwhite residents still struggle to get by.

When I pulled up to Los Cuates for lunch with Jimmy Santiago Baca, I couldn't help but reflect on the familiarity of the scene. The sun was out, as it nearly always is, and right across Lomas was Mr. Tux, the Jazzercise place, and Clark's Pet Emporium. Los Cuates itself had been going strong since the 1980s. Even as I enjoyed the nostalgic pleasure of seeing all the familiar businesses, it was soured by a creeping ennui. Albuquerque never fully recovered from the 2008 financial crisis: its population has plateaued, and the city is still struggling to reorient itself to the new normal after a century of growth driven by real estate development. This means that the bulk of Albuquerque, conceived as an American anyplace, remained in that generic mold despite the efforts of contemporary leaders to confront the conflicts of the twentieth century. There's a fine line between continuity and calcification, and I couldn't quiet the fear that no amount of righteous effort to rectify the iniquities of the past would be enough for Burqueños to fully throw off the weight of the city's history.

My misgivings faded as soon as I saw Baca, his face deeply lined under a Carhartt baseball cap. He's older now, closer to a literary dignitary than the dude who once crashed a bookseller's convention in Orange County rocking denim cutoffs in a room full of slacks and nylon stockings. As we got settled at a table under one of the restaurant's ornate vigas, he wanted to talk about his kids. "I couldn't be happier," he said, ticking through their various accomplishments: an award-winning filmmaker, a student at the University of Miami. "I have to give all of them money—I'm broke, but I'm okay with that. None of them are sick, they're all struggling to make it, and they're doing great."

As we chatted over enchiladas and tamales, I remembered first reading *Martín and Meditations on the South Valley* in high school. Baca's poetry had electrified me: it felt as if the entire city had been condensed into this one, slim volume. Even as most of the events of the book unspooled literal miles away from the stultifying suburbia I couldn't wait to escape, each line provided a new jolt of recognition. It was the first time I understood my hometown not as a forgettable backwater, but as a place with all the complications and contradictions necessary to fuel a work of genuine literature.

Eventually, I got around to asking Baca why it was that Chicanismos had faded from the culture, with so many Mexican Americans now preferring either more generic terms like Hispanic, or else a controversial neologism like Latinx. "I'm not an intellectual, man, I'm a writer," he replied. But after thinking it over for a few minutes, he assessed the shift. "Chicanos pretty much came out as political, recently escaped from the prison of

their own nomenclature. They didn't know what to call themselves until the mid-'50s, until the *Chicano* term came out and it
coalesced with political consciousness"—informed by police violence, the struggle to organize farm laborers, and opposition to
the Vietnam War.

But as Mexican Americans became more and more assimilated into mainstream life, the political salience of Chicanismos
became a liability. When he was a congressman representing
Santa Fe and the rest of northern New Mexico in the 1990s,
Bill Richardson remarked that the word *Chicano* had "a bad connotation" and that people of Latin American descent "can open
more doors with *Hispanic*." When it came to literature, Baca said,
"white editors in New York needed something like Maya Angelou" to sell to the growing class of self-identified Hispanics. This
made writers like Sandra Cisneros particularly appealing, given
that her books, most notably *The House on Mango Street*, are not
overtly political but instead preoccupied with familial striving. "I
love Sandra," Baca said, "but she began to rely more and more
on the directions of her editors in New York, rather than paying
attention to what was going on in the barrio. She had to wear the
mantle of the politically correct Latina."

For his part, Baca has resisted the trend of American literature toward attempting to mortar over the broader cultural frac-
tures that characterized his youth by hiring more people of color
into positions of power. Baca's star was at its greatest height in
the late 1980s, when he won an American Book Award and sold
the script for the film that became *Blood In, Blood Out*. Numerous universities offered him teaching jobs, but he turned them all
down, preferring to live simply in Burque and focus on his craft.

Baca's refusal to participate in academia (so rare among contemporary writers) is partially explained by one of his poems,
"My Dog Barks." The focus is a dehumanizing conference about
"writing / in prison," where the participants "prefer to translate
their suffering into MFA / papers." Most cuttingly, his speaker
wonders, "if they weren't in / prison / you wouldn't be able to
have a conference, / would you?"

Baca ordered another margarita, then turned ruminative.
"Books, for me, are everything." He recalled when he was first

teaching himself to read in prison, an experience that defied the identity-based politics that have become pronounced in contemporary American literature, where writers are routinely admonished for creating characters that overstep the bounds of their personal experience. "Turgenev was one of the first ones I read," Baca remembered, "and Emerson was another one. Wordsworth was another one. I wasn't thinking, were they Chicanos or ex-prisoners? They spoke to my heart. This guy going duck hunting in the morning in Russia, Turgenev? That's *me*, dude. That's me."

He painted the sort of mental scene that would play out in his cell in Arizona. "I brought Turgenev to myself. We're sitting there having Russian tea—a sip of vodka—talking about, 'You think there's birds in that bush over there?' Turgenev said, 'I don't know, let's have some more vodka and find out.'" Baca chuckled, as if describing a leisurely afternoon with an old friend.

I was struck by the way Baca still fiercely advocated for a version of Mexican American identity grounded in a politics of resistance, while at the same time locating himself within the European canon. He had effectively cut himself off from the award circuit that many writers of his generation and distinction are beholden to, but in a way that allowed him to retain a more intimate experience of the work itself. It was a balance that suggested a path toward a multiculturalism characterized by mutual respect instead of assimilation or identitarian battle. "It's been a really hard journey," Baca allowed. "With a lot of sorrow, a lot of fear. But all of it is not commensurate with the joy of writing a poem."

PART III

OUT OF MANY, UNO

Maryvale—Juárez—Henderson—Buckeye

VIABILITY

The "street" I lived on (it had been paved once, but now it was mostly broken asphalt and big pools of sand) ran north–south, and the house faced west, the direction the wind blew from, pecking the living room's picture windows with sand. A sign nearby read, "Next services 100 miles."

—RUBÉN MARTÍNEZ

Early in the evening along Phoenix's Palm Lane, the sprinklers switched on. The residential street cuts through an area called Encanto Park, extending from the state fairgrounds over Central Avenue and then into the next-door neighborhood of Coronado. As I strolled in the warm, February twilight under the regular procession of palm trees lining the street, so much water was shooting out over the grassy lawns that the cracks in the sidewalk became streams of runoff. Dodging rivulets, I paused for a moment on a dry patch of cement to look over the hedge of a Tudor-style mansion, in front of which a baroque fountain burbled to itself.

A mile due south, on the other side of Van Buren Street, there are no palm trees, no lawns, no hedges. In fact, once you get beyond the mid-rise office and condo buildings downtown, there are barely any houses. Many of the lots in South Phoenix are simply vacant expanses of pinkish gravel that separate the handful of workaday houses that remain in one of the city's original neighborhoods.

South Phoenix was formed after the railroad was first extended to Phoenix in 1887, creating a zone sandwiched between the tracks and the Salt River. The area was initially populated by Mexican farmworkers, but it gradually developed sizable Black and Indigenous communities as well. In the 1930s, the federal government's newly created Home Owners' Loan Corporation started publishing maps that color-coded neighborhoods to indicate their supposed level of mortgage risk. All of South Phoenix was redlined by HOLC as "hazardous," while most of the areas extending north, toward the Phoenix Country Club, were designated as either "best" or "still desirable."

The redlining of South Phoenix was a powerful incentive for couples like Lincoln and Eleanor Ragsdale to try to find a home on the other side of Van Buren. Lincoln had grown up in Oklahoma, the son of a mortician named Hartwell who operated a funeral home in the Greenwood District of Tulsa during the 1910s, around the same time Booker T. Washington labeled the neighborhood "Black Wall Street" in recognition of its status as a rare beacon of Black prosperity in the Jim Crow South. That light was extinguished in 1921, when a fabricated story about a white woman being assaulted by a Black man became a pretext for white Oklahomans to level the neighborhood. In just twenty-four hours, a mob lynched hundreds of Greenwood residents and immolated 1,250 homes and business. Hartwell Ragsdale escaped by secreting himself into a train car that rolled out of Tulsa as smoke billowed up around the tracks.

Lincoln was born in 1926 and grew up in the town of Ardmore, midway between Oklahoma City and Dallas. He enlisted with the Tuskegee Airmen immediately after graduating from high school in 1944, later telling the historian Matthew Whitaker, "I wanted to be a pilot because I wanted to prove something. We were treated as second-class citizens, but the only way to change something is to prove that you can do something." After finishing basic training in Mississippi, Ragsdale was assigned to Luke Air Force Base in Phoenix, one of the numerous airfields and other military facilities across the Southwest that had been expanded during World War II. Ragsdale left the service once the war was over, deciding to preserve the family tradition by opening his own mortuary in Phoenix, where white-owned funeral homes rarely worked with Black or Mexican American families.

Ragsdale managed to buy some land in South Phoenix with savings from his years in the Air Force, but no banker would underwrite the loan of $35,000 it would take to build a mortuary. That included Valley National's Walter Bimson, who had just provided $600,000 to Bugsy Siegel so he could finish the Flamingo, and then forked over another $300,000 to cover the enormous losses the casino absorbed in its first two weeks. Comparatively, a modest business headed by Ragsdale, a veteran from a family of successful entrepreneurs, was considered a nonstarter. Municipal officials were no more concerned with the plight of South Phoenicians than the bankers were—in the years after World War II, more than half of homes had no running water or bathrooms. Even in the late '60s, the area was described by *The New Republic* as a "squalid slum" where "streets are unpaved, sewers unconnected, and public utilities inadequate; many houses have outdoor toilets, and many houses are no better than outhouses."

Instead of traditional financing, Ragsdale finagled a personal loan from an architect named H. Harry Herrscher, a Greek immigrant who was sympathetic because he had been cared for by a Black teacher when he first arrived in America. The Chapel in the Valley opened in early 1948 and was swiftly dubbed "Arizona's Finest Negro Mortuary" by a local newspaper. Shortly afterward, Lincoln married Eleanor Dickey, a teacher from Pennsylvania. The couple moved into an apartment adjacent to the mortuary while they saved to build a duplex across the street. By 1953, their second child was on the way and business was good enough that the time seemed right to move on up. The Ragsdales set their sights on Encanto Park.

At the time, both Encanto Park and the adjacent Palmcroft District were whites-only communities. The latter had been established in the 1920s by Dwight Heard, the Chicago-born land speculator who worked with Del Webb, Walter Bimson, the Goldwaters, and the rest of the first wave of Phoenician boosters to literally put the city on the map. Heard built himself a house in the area, which was at a remove from both the Salt River and

the Grand Canal, and thus had few farms and little of the attendant dust and noise. The neighborhood Heard planned for the
eighty acres around his house was called Palmcroft, and in the
terms of sale for the other homes that spread out along the palm
tree–lined lanes, he included a covenant that prevented "those
having perceptible strains of Mexican, Spanish, Asiatic, Negro or
Indian blood" from taking up residence. Even as he was refusing
to allow Indians to be his neighbors, the businessman opened
the first iteration of the Heard Museum in a gallery next to his
house, which served to showcase his collection of Indigenous
artworks.

Once Palmcroft was built out, Lloyd Larkin and George
Peter developed the next block of land north of Palm Lane into
homes that would surround Encanto Park. They included a similar covenant, following Heard's lead in trading on the mystique
of Indigeneity while refusing actual Indians. Their model home
at 2040 Encanto Drive was built not with the bricks that characterized Palmcroft, but rather in the Pueblo Revival style popu-
larized by the New Mexico architect John Gaw Meem. It was
dubbed the "Indian house."

Not long after marrying Lincoln, Eleanor Ragsdale had
decided to learn the real estate business after she realized just how
difficult it was for her fellow Black educators to secure housing.
She knew that no white realtor would willingly sell a home near
Encanto Park to a Black couple, but since she had her own real
estate license, she was able to find out what homes in the neighborhood were on the market. With a complexion light enough to
pass as white, Eleanor attended open houses without any problem. She settled on a classic ranch house at 1606 West Thomas
Road with stucco walls, right across from a park. To circumvent
the covenant, Eleanor had a white friend make the purchase and
then transfer the house to her while it was still in escrow.

When the Ragsdales moved in, the community revolted. They
were bombarded by threatening phone calls, harassed by police
officers, and received a visit from a committee of neighbors offering to buy the house since they surely "wouldn't be happy here."
After the couple refused, the *n*-word was spray-painted across

the front of their house in two-foot-tall letters, which Lincoln refused to clean up, because he "wanted to make sure the white folks knew where the nigga lived."

After the uproar died down, Eleanor helped other Black families integrate neighborhoods across North Phoenix, all of whom received the same sort of welcome. Despite those efforts, the explicitly racist housing practices of the period resulted in decades of iniquity. In 1970, only 49 percent of Black Arizonans owned a home; by 1990, the proportion had dropped to 41 percent. In 2019, Black homeownership in the state was under 35 percent.

In part, the marginalization of aspiring Black homeowners was fed by the physical expansion of the Valley of the Sun, which caused a chain reaction of disinvestment in the neighborhoods closest to the city center. Maryvale, for example, once represented the western frontier of Phoenix, a wide expanse of onion fields along the Grand Canal. In the late 1940s, when urban Phoenix ended around 35th Avenue, a native Phoenician and World War II veteran named John F. Long built himself a home in a grove of citrus trees along Glendale Avenue, using a GI loan underwritten by Valley National. Before Long had even finished the house, he'd received numerous offers to buy it, some for more than twice the cost of construction. Recognizing the opportunity, Long built nine houses, then fifteen, then three hundred, reinvesting his profits into larger construction crews whom he trained in the nascent art of residential mass production. In 1954, he appealed to Walter Bimson for financing on his biggest bet yet: the acquisition of hundreds of acres west of the city limits, where he would lay out a massive, planned community. Bimson looked over Long's financial statement, commenting that he wasn't showing quite enough profit to justify the sizable loan he was after. Nevertheless, Long remembered, "he felt I definitely had something on the ball. So, they made the loan."

Long called his planned community Maryvale, after his wife. By 1960, the community had grown from 143 residents to more than 65,000 and been annexed by the city of Phoenix. Only 88 of the new residents in Long's cookie-cutter houses were Black, and

Hispanics, Indians, and Asian Americans likewise had difficulty buying into the neighborhood because of its covenants.

But as swiftly as white families had descended on Maryvale, they began to move away again, this time to newer suburbs to the north and west. The trend accelerated as the city seized land in the barrios of South Phoenix, with plans to bulldoze homes and replace them with more freeway lanes or industrial lots. A predominantly Mexican American neighborhood known as the Golden Gate was almost entirely cleared to make way for an airport expansion, displacing six thousand families. By 1985, only one building remained in the old barrio: the Sacred Heart Church. Forty years later, it remains standing as a gravestone for the Golden Gate, an abandoned redbrick temple surrounded by a chain link fence and a flat plain of dust, airplanes constantly ascending and descending over its domed steeple.

Given that the homes in Maryvale were relatively afford-able and their racial covenants had been outlawed by the Fair Housing Act of 1968, the neighborhood became a destination for displaced Mexican Americans, including many from the Golden Gate. The number of Hispanics living in Maryvale doubled between 1970 and 1980. The demographic cross-fading acceler-ated in the 1980s, particularly after the city shut down a number of wells in the neighborhood after tests found that the ground-water was laced with trichloroethylene, a carcinogenic solvent that had leached into the water table from a nearby industrial zone. Some wells showed toxin levels 350 times higher than what was considered safe by the Environmental Protection Agency. In 1987, the facilities that had polluted West Phoenix's groundwa-ter were declared a Superfund site. (Luckily, the Central Arizona Project aqueduct had just reached Phoenix, so the city could fully shift from pumping the local aquifer to supplying houses in Maryvale with water from the Colorado River.)

White flight from Maryvale was exacerbated by the rise in crime that so often accompanies displacement. "When neigh-borhoods are disrupted like that, the people lose their sense of community, of security," Miguel Montiel, a social work profes-sor at Arizona State University, observed to a reporter from *The Arizona Republic* in 1979. "They lose the stuff that holds a community together—the reliance on one another, the social interactions, the dialogues."

While Maryvale was once a close cross-section of the broader city, by 2000, average incomes in the neighborhood were 26 percent lower than in the rest of Phoenix, while educational attainment was also declining. Though many original inhabitants of Maryvale had left, they didn't necessarily sell their houses: once a community where most everyone owned the building they lived in, in the twenty-first century only about half of homes in Maryvale are owner-occupied. For many newer, Mexican American residents, this means that instead of building equity in the neighborhood, they are merely paying rent to an Anglo landlord who long ago decamped for Glendale or Peoria.

As South Phoenix was emptying out and neighborhoods like Maryvale absorbed their previous residents, new subdivisions kept being built farther and farther into the desert frontier surrounding the city. Between 1990 and 2020, the number of miles traveled by Phoenicians doubled, to more than 124 million per day, much of it over the 467 miles of freeways that now overlay the entire Valley of the Sun like a net. All those extra miles translated into the largest rise in vehicle emissions for any large city in the country, with the amount of exhaust produced in Phoenix tripling over the past thirty years, contributing ever more greenhouse gas to the atmosphere as well as leaving the city with some of the worst air pollution in the country, its smog so pervasive it has been nicknamed "the brown cloud."

Though parts of downtown Phoenix have been revived by the construction boom prompted by the opening of a light-rail line down Central Avenue, the neighborhoods south of that corridor remain among the most impoverished in the state. Even the increasing intensity of Phoenix's summer heat is falling disproportionately on South Phoenix, where it's often 10 degrees hotter than in neighborhoods with more shade and water features. That gap mirrors a trend across the entire Southwest, where the poorest 10 percent of neighborhoods experience significantly higher temperatures than wealthy enclaves in the same cities. This leaves many poor southwesterners to turn up the air conditioning, even when they live in drafty manufactured homes, racking up monthly electricity bills of three hundred dollars or more. Such overreliance on air conditioning has dire consequences for the climate. Already, a quarter of all electricity generated in Arizona supplies air conditioning units, four times the national average. Given that only 17 percent of the state's electricity is generated by renewable sources, that means Arizonans are adding millions of tons of carbon into the atmosphere every year just to survive summer temperatures that are only getting more lethal.

By orienting residential development and transportation planning around white supremacy, the early boosters of the Southwest created cities that have continued to enforce their preferred racial hierarchy into the twenty-first century. Though perhaps no longer motivated by explicit animus, any perpetuation of the same style of real estate development only makes the gap wider—as subdivisions sprout further and further afield, more and more people are drawn out of historic neighborhoods, straining civic resources and generating an ever-greater volume of emissions. It's possible to correct that legacy, to reimagine cities across the Southwest as places where rising temperatures and dwindling water reserves don't have to culminate in a great climate migration away from the arid lands. Doing so, though, will require listening not just to the lessons of the modernist architects who pioneered sustainable building practices, but also to traditions that stretch much further back in history, to when the region's earliest residents understood that only through banding together would it become possible to thrive in the desert.

The Southwest's most famous house is a seasonal residence. Taliesin West sits on an overlook in the McDowell Mountains north of Scottsdale, an alluring adaptation of the low, spacious homes Frank Lloyd Wright pioneered in the Midwest to what he perceived as the more rapturous, angular spaces of the Sonoran Desert. Seeking to evoke the jaggedness of peaks like Camelback Mountain and the thread-sharp prickers of a cactus, as well as elicit the play of light and shadow that is everywhere apparent in the southwestern landscape, Wright built the walls of Taliesin West out of stones from the hillside and had his apprentices imprint triangular grooves along their length. Rather than rely on the clean, right angles he favored for his prairie homes, Wright set every vertical plane at a 15-degree angle, skewing them just enough to create a strikingly disjunctive appearance.

These decisions proceeded from what Wright called his "prescription for a modern house: first a good site . . . then build your house so that you may still look from where you stood upon all that charmed you and lose nothing of what you saw before the house was built, but see more." Beyond merely designing for the resident, Wright also believed that "any building for humane purposes should be an elemental, sympathetic feature of the ground, complementary to its natural environment."

For all his attunement to the geography of Taliesin West, Wright was oddly unconcerned with the climate. The house was oriented on a northwest-southeast axis, allowing a panoramic vantage of the Valley of the Sun at the same time it guaranteed more heat exposure than if it had simply faced south. The walls that were not made of foraged rock were fashioned of glass; so were the ceilings, albeit covered over with fabric to diffuse light. All the glass creates a lovely effect in the winter, maximizing sunlight when the weather is mild and the days are short. But in the summer, Taliesin West is a sauna. Not that Wright was ever there for the heat: from May until October, he and his wife returned to their home in Wisconsin.

After Wright's death in 1959, Taliesin West was transformed into a museum, meaning that it was necessary to tweak the building to make it habitable year-round. In the 1990s, the carpet of the long, lovely living room had a divot cut into it to make way for an A/C vent. Even the complex's small movie theater, the only room with no windows, now includes the sort of climate control unit you can buy for a hundred bucks at Home Depot—and a good thing, too, because the homebrew concrete that space is made from would otherwise operate like the walls of an oven when the mercury inevitably surged into the triple digits.

The tour guide paused on the luxurious green lawn outside Taliesin West's main living room and pointed to the power lines that cut diagonally across the view, near enough that their steel towers appeared taller than the distant Phoenix Mountains.

Shaking her head in dismay, she related how heartbroken Wright had been after the lines went up in the 1940s—the architect even went so far as to write a personal appeal to President Harry Truman asking for his assistance in having them buried. The small tour group groaned in resignation, agreeing that it was a shame.

Frank Lloyd Wright's distress and the contemporary crowd's sympathy provided a neat encapsulation of the studied blindness of the colonizer. Those power lines were originally built to connect the city to the Hoover Dam, which generated a tremendous amount of electricity by turning 120 miles of the wild Colorado River into Lake Mead. As time wore on, the same lines allowed Phoenix to draw power from plants burning coal scraped off the Colorado Plateau by Hopi and Diné miners, as well as the Glen Canyon Dam after it was completed in 1963. For Anglos to find the Valley of the Sun comfortable they needed air conditioning, but the interruption of their scenic view by a piece of infrastructure that might force them to consider the enormous network of extractive industry and ecological disruption that made their comfort possible—such an intrusion was too much to bear.

Seeking a different perspective on building for the desert, I reached out to Jesús Robles and Cade Hayes, the cofounders of a Tucson architecture firm called Dust. Hayes grew up in Carlsbad, not far from the spectacular caverns in southern New Mexico, while Robles is a native Arizonan. The pair met in West Texas as architecture students at Texas Tech. Despite Lubbock's associations with the oil and gas industry, Hayes referred to the area as Comancheria—the heart of the region the Comanche dominated for more than a century.

That attunement to the Indigenous history of the borderlands that extend from West Texas to southern Arizona is typical of Robles and Hayes's practice. Once, a client came to their office and said, "You guys are desert architects? I want this," handing them an image of a house in Santa Fe that she'd ripped out of a magazine: adobe walls studded with small windows and vigas, the unrefined logs thick enough to support the weight of a flat roof. Robles told her, "'Well, that's appropriate for Santa Fe, but maybe not for Arizona.' She didn't know what to do about that."

Hayes nodded. "People's perception coming from Minnesota is that it's all the same. 'I want some coyotes and Kokopellis.'"

Instead of abiding by an idea of southwestern architecture that emphasizes familiar styles, Dust is grounded in the peculiar features of whatever ecology Robles and Hayes are commissioned to build within. For a house called Casa Caldera in the San Rafael Valley, which bridges the U.S.-Mexico border near Nogales, Hayes said, "We were fortunate to have a client say, 'I don't want air conditioning.' That puts your feet to the fire. You have to design a building that's efficient, can hold heat but also withstand heat. And can be fully ventilated." With all that in mind, the architects spent as much time on the site as possible, just as Judith Chafee had when she began work on the Viewpoint House. "That informed our decisions and designs," Robles said. "It's a grassland, it's a little higher up, it has a wider range of hot and cold, the landscape can regenerate quicker. These are all new sensibilities for us to process and respond to."

The result was a small, off-grid house that mimics the zaguáns that became common throughout northern Nueva España during the colonial era. Zaguáns have an open, central corridor that acts as a breezeway, cooling the rooms by moving air through them. The primary building material at Casa Caldera is a mixture of volcanic rock and cement that's been rammed into form. At less than a thousand square feet, it's a residence with some humility to it, meant more to settle into the landscape than predominate. Crucially, there is no air conditioning. As far as the natural climate control goes, Hayes said, "I think for some people it can be a little hot in the summer." He shrugged. "But that's how we lived before the advent of air conditioning."

What Robles and Hayes were describing was a bid to infuse the traditional form of a zaguán with a contemporary sensibility, one that worked within the established limitations of the climate rather than seeking to overcome them with technology. This approach reminded me of an essay written by John Gaw Meem in 1934 that sought to explain the approach behind his Pueblo Revival buildings in Albuquerque and Santa Fe. "Some old forms are so honest, so completely logical and native to the environment that one finds—to one's delight and surprise—that modern problems can be solved, and are best solved, by use of forms based on tradition," Meem wrote. "It is truly refreshing to feel that in our contemporary architectural movement is still an opportunity for the expression of ancient values."

For now, the sustainable architecture of Robles and Hayes, like that of Chafee and Meem before them, remains available only to a select few. Still, the attitude they bring to building in the desert has become steadily more popular over the past few decades, with sustainability itself becoming a buzzword embraced by the region's latter-day boosters.

In 2009, the mayor of Phoenix, Phil Gordon, unveiled a seventeen-point plan for greening the city that included everything from building the Valley's first solar plant to retrofitting businesses along the city's lonely light-rail line to make them more energy efficient. A decade later, his successor, Kate Gallego, embraced a proposal to expand that light-rail into a genuine network, with the goal of adding forty-two miles of new service by 2050. Gallego also aimed to repave streets with heat-retarding chemicals, reduce Phoenix's municipal waste and emissions, and partner with utilities to incentivize EV ownership. The goal, she wrote in 2020, was "making Phoenix the most sustainable desert city on Earth."

The most vital component of fulfilling that ambition—and the one with the longest track record of success—is a transformation of the city's relationship to water. In 1990, per capita use in the city was 250 gallons a day, far outpacing the national average of 160 gallons. Over the following thirty years, the national average remained roughly the same while the daily water consumption of Phoenicians declined by almost 30 percent, with the two numbers converging around 2020.

The biggest driver of conservation in Phoenix was a simple policy reform: charging more for water in the summer. Kathryn Sorensen, a former head of the city's water service, told me, "That one action probably saved more water in Phoenix than all subsequent actions combined. If you want to have a lawn in the summer, it's going to take this vast amount of water. And then if you charge *more* for water during that time, people learn quickly,

'I like my lawn, but I don't like it four hundred dollars a month worth.'" Though households in Phoenix typically use twice as much water in the summer as they do in the winter, under the original system they paid the same flat rate year-round. In 1990, the city took out an ad in *The Arizona Republic* to explain the new rate structure, promising that "water bills, on a yearly basis, will decrease for nearly half of all residential customers"—savings that would be subsidized by the most profligate users.

Sorensen grew up in Tempe. While finishing her dissertation in agricultural economics at Texas A&M, she returned home to visit her father and went for a hike up Piestawa Peak, which overlooks the north side of Phoenix proper. "At the top," Sorensen said, "you can see various Salt River Project canals, and you can see the pattern of growth across the city. You see the neighborhoods, the suburbs, the high-rises, the golf courses, the resorts— the whole pattern of development. I remember standing on the top of the mountain and saying to myself, 'Someday I'm going to be in charge of this.' Like most twentysomethings, you don't have a lot of self-confidence, you don't really know what your future holds. But it was just elemental."

Her first crack at that vision was in Mesa's water department, where she spent the first decade of her career—a meaningful appointment, given that Sorensen is a descendant of one of the first families sent by Brigham Young to settle Mesa at the end of the nineteenth century. She rose to become director of the department, before jumping to the top job in Phoenix in 2013. By that point the free market approach to conservation had already worked wonders, with average household water consumption in July dropping from 750 gallons a day to just five hundred. But those gains in efficiency were threatened by the city's continuing expansion, which was accelerating again after the short layoff provoked by the 2008 housing crisis.

"We didn't change the structure, but we kept increasing the rate," Sorensen said. "You're willing to pay four hundred dollars a month for a lawn, how about five hundred?" She paired the blunt force of rate increases with programs aimed at changing the city's culture around water use for the long term. "We really emphasized giving people the tools they needed to make informed choices, being out in communities and neighborhoods, engaging with them, with a goal toward helping people understand that, if you're going to live in the desert, you're just going to need to live a little differently than what you might have been used to."

That approach is a stark departure from what Sorensen called the "command and control programs" that have grown in popularity across the United States, which feature temporary restrictions around certain types of consumption. In Sorensen's eyes, repeated calls to cut consumption were counterproductive, since they signaled that conservation was a short-term ask rather than a lasting reform. Make that ask too many times, and "it becomes nonsensical, like crying wolf. People just tune you out." To her point, after Governor Gavin Newsom asked Californians to cut their water use by 15 percent in 2022, consumption actually went up. "When you work on changing the culture, that can become a more permanent thing."

The success of Phoenix's quiet pivot to conservation in 1990 has been remarkable. Since then, the city has added seven hundred thousand residents while its water consumption has remained stable. Still, Sorensen's faith in the free market was typical of someone who has spent their career inside the Salt River Valley growth machine, and I couldn't help but recognize the inadequacy of price incentives when it came to taming the waste of the wealthiest Phoenicians. Those willing to pay an exorbitant water bill could use as much as they wanted—as demonstrated by the overwatering I observed along Palm Lane.

A more powerful reform would be to prioritize dense, infill development over new subdivisions. Exurban growth doubles the impact of water shortage crises on cities, in terms of both magnitude and duration. At the same time, allowing development far from urban cores forces residents into long commutes that translate into outsize carbon emissions, which in turn exacerbate aridification.

"Land use patterns, that's our next frontier in terms of water conservation," Sorensen told me. "Within urban areas, use varies tremendously depending on whether you're building a subdivision with large, expansive lots versus a very high-density, multistory development." The main source of water for the city of Phoenix is the Salt River Project, the contemporary version of the agency that first began delivering water to the city from the Salt and the Verde in the early twentieth century. "Because it's converted from an agricultural to urban system over time, the Salt River Project serves half the water that it used to," Sorensen said. The Arizona Department of Water Resources estimates that one acre-foot of water is enough to supply three households for a year, while growing an acre of cotton can require four acre-feet of water. Put another way, one acre of cotton requires the same amount of water to support as sixteen homes. If Phoenix's housing stock shifted toward apartment buildings instead of ranch houses, that ratio would only become more dramatic.

In recent years, Lincoln and Eleanor Ragsdale's original neighborhood in South Phoenix has been rejuvenated. That process began in the 1990s, when artists started populating the area around Roosevelt Street, forsaken for decades. The so-called Roosevelt Row scene was anchored by small galleries like the MonOrchid and Holga's, as well as underground performance venues like Metropophobia and the Planet Earth Theatre. On their bare-bones stages, a handful of likeminded members of Phoenix's convivial avant-garde could watch an experimental duo known as Sappho's Fist use a bit of wire to scrape a bowl directly in front of a microphone, or Christopher Marlowe's *Edward II*, as performed by an entirely nude cast.

In 1993, the artists from the Roosevelt Row galleries began to organize a monthly open-studio series, called First Friday. Those events evolved into an all-out urban festival, what the artist Nan Ellin once described as a "party of plein aire fashion shows, performance and installation art, belly dancers, hip hop troupes, 'U-Haul galleries' rented for the evening and parked on vacant lots, and an itinerant band called the MadCaPs who play from the bed of a roving pickup truck."

Thirty years on, downtown Phoenix's rowdy past is all but invisible. Though the area has life again, the neglected neighborhood of yesteryear is less a thriving urban enclave than a portrait of the real estate market run amok. Roosevelt Row now alternates between parking lots and glassy mid-rises whose ground floors are occupied by restaurants with faux rebellious names—Breakfast Bitch, Pour Bastards, the Killer Whale Sex Club. Modified Arts and Eye Lounge are the only galleries still hanging on amid the yuppie dreck. The abundance of parking undercuts the appeal of the light-rail on Central for anyone who wants to explore the neighborhood, meaning that the majority of pedestrians circulating around downtown Phoenix are people who have been made homeless by the city's metastasizing housing crisis.

Despite the phenomenal number of houses that are built in the Valley of the Sun every year, need for shelter still outpaces demand. The average price of a home more than doubled between 2000 and 2020, while salaries only rose by around 50 percent; even as thousands of apartments were being built downtown during that period, the average rent of those units was significantly higher than elsewhere in the city. By the end of 2019, the waiting list for a Section 8 subsidized housing voucher had surpassed sixty-two thousand families, with an anticipated wait time of up to ten years.

When the pandemic hit, many of those who were already struggling ended up on the street. In 2021, around a thousand homeless Phoenicians started congregating across a few blocks of downtown Phoenix, an area of camping tents and improvised structures that became known as "the Zone." Over the next year, sixteen people in the Zone died from a mixture of hypothermia, heat exposure, and drug overdoses. Eli Saslow of *The New York Times* interviewed a woman named Shina Sepulveda as she attempted to disinfect one of her neighbor's stab wounds. "This isn't a life," she said.

The homeless population of Maricopa County approached ten thousand in 2023—32 percent higher than before the pandemic. In response to a lawsuit from business owners, a federal judge ordered the city to clear the Zone, dispersing its population to a variety of shelters and a hastily set up safe-camping site— known as the "SOS lot"—at 15th Avenue and Jackson, where the city purchased an old warehouse so it could offer air conditioning to the three hundred people claiming a portion of the massive parking lot outside.

The swiftness with which Phoenix was able to put together the SOS lot demonstrates just how much of the city's fabric is underutilized. There are twenty-four thousand acres of completely undeveloped land within Phoenix, and that's not even accounting for other defunct warehouses or vacant strip malls. At the city's current density, that land could provide homes for 118,000 people. If Phoenix instead changed its zoning code to allow density akin to Los Angeles, there could be enough room for more than three hundred thousand new residents, all on land that is currently bare.

It's ludicrous for developers to build new subdivisions thirty miles from downtown Phoenix while neighborhoods that are little more than fifty years old degenerate into a patchwork of clearings. City officials have advanced plans for reviving a 690-acre swath of Maryvale that it dubbed the "village core," imagining all the parking lots that surrounded a defunct mall as a walkable district of mid-rise apartment buildings and green space, connected to the city's center by an expansion of the light-rail. Likewise, the abandoned church in the Golden Gate, near the Phoenix airport, could readily be converted into an apartment complex. There's no shortage of popular appetite or economic need for the kind of dense infill that might allow Phoenix to truly become the "most sustainable desert city on Earth." What's lacking is the will to upend the century-long tradition of desert-eating development that created today's iniquitous metropolis.

The desert is not antithetical to human civilization. Quite the opposite, arid regions have always served as a crucible for forging societies. But living there in perpetuity means adopting an outlook that accepts that there are limits and adjusts accordingly. Over the next century, Earth will get hotter, and more places will look like the Southwest; meanwhile, the Southwest's summers will become more extreme, pushing the limits of human physiology to the breaking point. Between 2019 and 2023, more than eighteen hundred people in Maricopa County were killed by the heat. That surge parallels a rise in heat-related deaths nationwide, the toll exacerbated by a pervasive homelessness crisis that exposes more than 180,000 people to the elements every year. More and more Americans will be killed by future heat waves

until cities abandon development patterns that exclusively consider the comfort of the wealthiest individuals in favor of a more
communal style of living, one oriented toward the good of the
whole.

How different would the contemporary Southwest be if, when
Anglos arrived, they'd simply accepted that the desert was hot?
If they'd looked around, seen that the existing population was
already adapted to the conditions, and followed their lead? In an
interview, the German industrial designer Dieter Rams remarked
that "architecture, design, even the way communication is organized, are expressions of the socio-political reality, and yet can
also affect and shape this reality." The urban fabric of the Southwest was transformed over the past century, but that does not
mean its current form is final. The lessons for how to survive a
sweltering world have already been taught: to survive the twenty-
first century, we'll have to return to traditions of cohabitation and
fellowship, fostering an interdependent community in recognition of the inherent limits of any place we hope to call home.

Driving between Las Cruces and Alamogordo, one shoulder of Route 70 looked like the rest of southern New Mexico: a predictable expanse of gramma grass broken up by the occasional sage-colored creosote bush or the high stalk of a soaptree yucca. On the other side of the road, a ridge of granulated gypsum formed waves of hills, their blinding white color only slightly tempered by rosemary mint and skunkbush sumac, with roots riven through the sand to seize whatever moisture they could. The gypsum that created White Sands National Park had been deposited hundreds of millions ago, when nearly all of today's Chihuahuan Desert was submerged under the Permian Sea. Once the basin between the San Andres and Sacramento Mountains went dry, the wind channeled up the corridor between the two ridges broke the mineral down into finer and finer particles, until only brilliant grit remained.

Distracted by White Sands, I didn't realize I was nearing a Border Patrol checkpoint until I saw the red brake lights of the handful of cars that had been diverted from the state highway and into a queue under a towering metal shade structure. When it was my turn I rolled down the window and the agent took a quick look at me, his arm draped over an AR-15. "Good afternoon, sir. Are you a citizen?" I said yes. "All right, have a good day!" I'd hardly taken my foot off the gas, as at all the other checkpoints I'd passed through in recent years in the Imperial Valley, West Texas, and southern Arizona. When you're Anglo and spending a lot of time in the borderlands, these micro-interactions become routine.

After a stop for coffee in Alamogordo, I headed back toward El Paso on Route 54. Once the Sacramentos were in my rear view, the Chihuahuan Desert took over, little different here than the stretch between Van Horn and Marfa, or even between Juárez and Ascención. The monotony of these drives is harder to bear than those of comparable length up north, through the spectacular Painted Desert of Arizona or the mesa country of western New Mexico. Luckily, the sky offered up a bit of action. Dark clouds approached from the southwest, probably one of those systems that starts in the Sea of Cortez and then cranks east all the way to the Great Plains. Again, the road split: to the right, a mammoth bank of cerulean vapor; to the left, an untroubled sky of periwinkle blue. As the ethereal trails of virga touched down around the horizon, I rolled down the windows to try to catch the petrichor, maybe hear some thunder. The southern clouds were getting darker and darker, even as the sky out to the east remained clear.

I kept waiting for a flash of lightning for the rain to come fast and hard and give the thirsty earth a drink. Instead, the system blew through. By the time I crossed the Texas line, the only clouds I could see were fluffy and white, the sunlight pervasive. With the staccato silhouette of the Franklin Mountains just coming into view, there was a great whoosh of air and metal as a train rushed over the tracks that paralleled the road. Its containers were stacked in pairs, the row of them extending for a mile as the engine heaved north.

MUTUALISM

And the dust that
confuses countries
is in your eyes, and you
blink, and you blink,

and you blink.

—ROSA ALCALÁ

In the summer of 1980, forty-three Salvadorans began their long journey to the United States by taking a bus to the city of Tapachula, in the southernmost corner of Mexico. Each of them had paid $1,200 to a pair of men named Carlos Rivera and Elias Nunez-Guardado, who operated what amounted to a travel agency for migrants out of a television repair shop in San Salvador. From Tapachula, Rivera and Nunez-Guardado chartered another bus to Mexico City, and then on to the northern state of Sonora, where the group finally disembarked in the border town of San Luis. Already 2,500 miles from the lush jungles of Central America, the Salvadorans sheltered from the dusty city's triple-digit heat while Rivera and Nunez-Guardado met up with two Mexican coyotes, who agreed to guide them across the desert on the way to the group's ultimate destination: Los Angeles. In California, three young women—two teenagers and a twenty-year-old—would be met by their mother, Rosa Huezo, who had paid their way and been misled to believe the group would be arriving by airplane.

The migrants were broken into two groups of sixteen, each

of them crowding into pickup trucks that drove east. Eleven of
the Salvadorans—mostly women and children—stayed behind in
San Luis, either because there wasn't enough room in the pickups
or they recognized just how underprepared they were to reach
America on foot. Just before daybreak, the remaining migrants
arrived at a restaurant in a village called Sonoyta, catching a little
shut-eye in the beds of the trucks until the restaurant opened and
they could all get breakfast. The Salvadorans spent the day in
Sonoyta, a few of the men accepting invitations from Rivera and
Nunez-Guardado to find a bit of liquid courage.

Despite Sonoyta's position in the heart of the Altar Desert,
where yellow dunes drift into peaks of sand hundreds of feet high,
many of the Salvadorans still believed their guides' assurances
that it would be a short walk to Los Angeles. Waiting out the heat
of the afternoon, a few lounged on bags packed with books and
keepsakes from home. A couple of the women didn't even think
to change out of their high heels.

Before Rivera and Nunez-Guardado ushered the group back
into the pickups at nightfall, they purchased twenty gallons of
water from the restaurant. Then the trucks drove about six miles
back to the west, parallel to the international frontier. They met
up with the Mexican coyotes not far from Quitobaquito, a small
pond fed by a natural spring in Organ Pipe National Monument
in the southwest corner of Arizona. After the coyotes showed
the Salvadorans how to crawl through a fence that followed the
road—a decades-old partition of barbed wire the National Park
Service had erected to prevent the spread of cattle-borne foot-
and-mouth disease—the group continued due north.

By ten the next morning, the water was gone. One man
pleaded for a break from the rigorous pace set by the coyotes; a
woman grew faint and had to be revived. As the migrants continued north, they kept shedding belongings and clothes. Over the
course of a few hours, a trail had formed: sweaters, pumps, photo
albums. Once the sun was directly overhead and the air temperature was approaching 105 degrees, the Salvadorans demanded a
rest. Several people collapsed in a dry arroyo shaded by paloverde
trees, hyperventilating, while a younger man doubled back to
round up a few stragglers. The older gentleman who had asked to

rest in the morning was nowhere to be found. One of the Mexicans gathered a few empty bottles and offered to fill them nearby, leaving with four migrants who volunteered to go along.

The Salvadorans who remained behind in the arroyo were so thirsty they broke off the branches of the paloverdes and tried to gnaw the supple bark, or pared the needles away from a prickly pear and chewed on the cactus's dense flesh. One of the women began pleading for someone to simply kill her, believing a swift death would be better than slowly perishing of thirst. Rivera lunged at her and had to be restrained by another woman, whom he slugged instead. The migrants emptied their packs, looking for anything that contained even a trace of moisture. They took sips from a cologne bottle and licked shaving cream. A few peed on rags and wiped their faces, hoping the urine would rehydrate their screaming skin. When the woman who had been attacked by Rivera came to, the last Mexican coyote had disappeared and four of the other women were dead. Rivera himself lingered nearby, lapsing in and out of consciousness.

Just a mile to the east, the Mexican coyote and two of the Salvadorans stumbled upon Route 85, which runs north-south through Organ Pipe National Monument. Two Border Patrol officers found them at the roadside that evening, but when they were questioned about who else they had been traveling with the three men refused to answer. Meanwhile, two other Salvadorans had set off from the shady arroyo on their own, a man and a woman. They kept walking through the night and the next morning, until they finally found the tiny town of Why—twenty-six miles north of where the migrants had entered the United States. After the pair found a gas station, the man ordered eight Cokes. Given the rough shape he was in, the attendant called the authorities; after the couple was picked up, they admitted to traveling with about thirty others.

Border Patrol agents found the first batch of marooned migrants about forty minutes later. Six men were sheltering in a gulch next to the corpses of two of their original companions. One of the officers, Hector Ochoa, poured much of his personal five-gallon jug of water over the bodies of the survivors, then soaked some of the clothing they'd stripped off their bodies, instructing them to suck on the damp fabric in order to gradually rehydrate themselves; had they attempted to drink water on its own, their bodies would have been so unprepared for the sudden rush of liquid they might have drowned. While that group was transported to safety, the rest of the Salvadorans were found under the paloverdes a short distance away—ten had died, with only three survivors lingering among the bodies. The casualties included all three of Rosa Huezo's daughters. "It is a terrible, terrible thing," she later told the reporter Louis Sahagún. "I wanted my daughters here because in El Salvador they are killing so many people. Now my children are dead."

In one of his sermons, the longtime head of Tucson's Southside Presbyterian Church, John Fife, described receiving a call for help from the town of Ajo, where the twenty-six rescued migrants were taken to recuperate. "The hospital asked some of us who were pastors to provide pastoral care for the survivors," Fife said. "We began to learn stories about death squads, and torture, and persecution of the church, and labor unions, and all those who were working with and on behalf of the poor in El Salvador."

As more and more Salvadorans were rescued from the desert of southern Arizona in the early 1980s, Fife and other local faith leaders educated themselves about the history of what the Salvadoran American journalist Roberto Lovato has called "the tiny country of titanic sorrows," from the genocidal campaign against its Indigenous population in the early twentieth century to the financial and military support the Reagan administration was providing to the junta that was killing tens of thousands of its own citizens in the '80s. Those same American officials insisted that the Salvadorans fleeing the violence were only crossing the border to find work, and thus did not qualify as refugees under international law. Appalled, the clergy embarked on a legal aid effort for migrants being held at detention centers near the border. "We would file political asylum applications for them," Fife remembered, "and we would bond out those who had been in detention the longest and bring them here to Tucson."

When *The Tucson Citizen* profiled John Fife in 1979, he was pictured in a denim jacket, goatee, and thick glasses, standing in front of the simple wooden cross in Southside's chapel, which was lined with metal folding chairs instead of pews. Southside had evolved into a radical congregation under the leadership of Fife's predecessor, Caspar Glenn, who pushed to desegregate Tucson during the 1950s and '60s. Fife was working to continue that tradition by fostering a congregation that actually reflected life in the city. The crowd at a typical Sunday service included "blond-haired girls in blue jeans and tennis shoes seated next to Mexican-Americans with goatees, *guayabera* shirts and slickly polished shoes, and with blacks, dressed in suits and ties, seated next to whites in plaid sport shirts." Fife told the paper, "I don't affirm an ideology. I want to correct the abuses of whatever system we have."

After an influential local activist named Jim Corbett began ferrying refugees across the border and hosting them in his apartment in Tucson, his wife approached Fife, asking him, "I've got twenty-one Salvadorans in my house—can we bring some of them to your church?" The answer was a resounding yes. After six months, the Border Patrol learned that Southside Presbyterian was harboring undocumented migrants and warned Fife that any parishioners who helped could face legal charges. Undaunted, the church went public with its campaign in early 1982, addressing letters to the U.S. attorney general and the heads of both the Border Patrol and the Immigration and Naturalization Service. "We expected to be indicted," Fife remembered, but the government decided to ignore Southside, recognizing that prosecuting church leadership could be perilous. "A movement began. The Sanctuary Movement. Southside church provided the model for other congregations, Protestant and Catholic and Jewish congregations, who were equally morally and ethically troubled by our government's policy here on the border."

Within two years, 287 churches had declared themselves as sanctuaries, and they were soon joined by more than a dozen cities. In an apparent effort to stem the movement's momentum, immigration officers arrested an activist in South Texas for transporting three Salvadoran refugees; in protest, Joseph Weizenbaum organized a Freedom Seder at Tucson's Temple Emanu-El and volunteers drove eighty Central Americans from Southside

Presbyterian to the synagogue in a two-mile convoy. In 1985, sixteen Sanctuary Movement participants were indicted by a federal
grand jury on charges of "alien smuggling," including Fife and
Corbett, along with Ramon Quiñones, a Catholic priest from
Nogales, and three nuns.

If federal officials thought the prosecutions would end the
Sanctuary Movement, they were wrong: over the course of the
trial, the number of participating congregations more than doubled. All eight of the activists who were eventually found guilty
were given probation by a judge who acknowledged that, though
they had broken the law, they had "done so because of humanitarian concerns." In 1987, Arizona Senator Dennis DeConcini
and Massachusetts Representative Joe Moakley introduced legislation to prohibit the deportation of Salvadoran and Nicaraguan
refugees already in the United States, citing the sixty thousand
Salvadorans who had died in that nation's civil war and the human
rights abuses of both the Sandinista government in Nicaragua
and the conservative, U.S.-backed Contras that opposed them.
Three years later, Congress created the designation of Temporary Protected Status for certain immigrants and extended it to
the half-million undocumented Salvadorans then living in the
country.

Over the previous decade, around thirteen thousand Central Americans had been hosted by Southside Presbyterian. "I
don't know how we provided hospitality to all those people," Fife
sermonized. "I know we didn't do it all by ourselves, it took a
whole community to do that." Though the Sanctuary Movement
had run its course, the steep rise in refugees traveling through
the deserts of southern Arizona throughout the 1990s kept the
church at the forefront of the struggle to support the undocumented. After the Clinton administration clamped down on the
two largest ports of entry during Operation Hold the Line in El
Paso and Operation Gatekeeper in San Diego, more and more
migrants were incentivized to try their luck along the six hundred
miles of desert separating the two cities. While apprehensions in
those urban sectors dropped by 70 percent, arrests in the region
south of Tucson rose by a factor of six.

In response, Fife helped found a group called Humane Borders in 2000, which established and maintained water stations across the Arizona borderlands. After fourteen Mexicans died attempting the crossing in 2001, Humane Borders held a funeral at the First Christian Church on Speedway Boulevard. The group's president, Robin Hoover, eulogized them by saying, "We are here because migrants are being systematically herded onto death trails." In 2004, another group, No More Deaths / No Más Muertes, joined the campaign, and since then, both groups' volunteers have played a high-stakes cat-and-mouse game with Border Patrol officers, who typically drain jugs of water whenever they find them.

In his memoir about working for the Border Patrol, Francisco Cantú wrote that the reason officers "slash their bottles and drain their water into the dry earth, that we dump their backpacks and pile their food and clothes to be crushed and pissed on and stepped over, strewn across the desert and set ablaze," was in hopes that migrants, once they "find their stockpiles ransacked and stripped," will realize "it's hopeless to continue . . . they'll save themselves and struggle toward the nearest highway or dirt road to flag down some passing agent."

Sometimes, the strategy works. Other times, it doesn't, and migrants stranded without food or water die. Between 2001 and 2021, the Border Patrol's budget expanded from $1.1 billion to $4.8 billion. Over the same period, an investigation from the *Arizona Daily Star*'s Curt Prendergast and Alex Devoid found that, while apprehensions plummeted by 90 percent in the Tucson sector, nearly ten times as many human remains were being found in the desert every year. "The journey has grown more dangerous over the years, due in part to walls and barriers funneling migrants into remote areas. Today, remains are found an average of seventeen miles from towns," Prendergast and Devoid wrote. "Exposure to the elements, particularly heat, is the most common cause of death, leading to 1,390 deaths and likely many of the 1,866 sets of skeletal remains where cause of death was undetermined."

Over the past forty years, a bipartisan desire to limit the number of undocumented migrants entering the United States has transformed the desertine borderlands into a fully militarized

space. Death has become, as a matter of policy, the ultimate deter-
rent. On the ground, though, the denizens of communities like
Tucson, El Paso, and Juárez understand that refugees will brave
any conditions if they believe it can help them forge a better life.
In cities where so many family histories reflect the same aspira-
tion, migration is understood not as a crisis but as a manifestation
of the natural striving of impoverished and dispossessed peoples
toward safety and opportunity. New arrivals do not imperil the
heritage of long-established communities—they are instead joining a pluralistic society already in full flower.

South of Paisano Drive in downtown El Paso, the streets were
lined with stores with names like Bodega de Cobertores, Orale
Mezclilla, and Todo Fashion. On the sidewalk, collared shirts
hung on plastic rolling racks and manikin legs advertised jeans. A
man in a backward baseball cap towed a cooler, shouting, "¡Pale-
tas!" in hopes some passersby might be looking for a taste of
relief from a sweltering June day, while ranchera guitar riffs and
reguetón beats pulsed out of the storefronts. Just before the point
where El Paso Street splits into a rotary directing traffic away
from the international bridge, there was a booth to exchange
pesos for dollars and a restaurant called Chelito's Gorditas, Bur-
ritos y Mas. Over on Santa Fe, buses queued up, bound for points
as distant as Idaho. I recognized the green-and-red branding of
Limousine Express, which connects to Los Angeles with stops at
a strip mall near I-25 in Albuquerque, a storefront on Jefferson
Street in Phoenix, and a parking lot next door to a boxing gym
in Las Vegas.

I made my way up the pedestrian bridge to Juárez, which was
covered by a sunshade and divided from the endless line of semitrucks waiting to get into the U.S. by a fence. On the right side
of the walkway, a metal screen perforated with thin ovals allowed
a limited view of the border itself. Just north of the rust-colored,
razor wire–topped wall, the American Canal channeled away
most of the Rio Grande's muddy water, while Mexico's share was
diverted into the Acequia Madre, leaving the Rio Grande itself

to run dry in the concrete trough that marks the official division point between El Paso and Juárez, El Norte and Latin America.

While only a few traces of graffiti were visible on the American slope of the canalized Rio Grande, the Mexican half was heavily illustrated. One panel looked like a language-learning flashcard, with the word "El Puente" printed neatly below three brown figures forming a human bridge across a river with their hands. Next to that image, an Arabic phrase was written against a geometric background, as if it were a mirror held up across the world, from one lethal border to another: فلسطين حرة ("Palestine Is Free"). As if in response, another mural read "Ni Muros, Ni Militares," a challenge to the militarization of both the border and Juárez itself, with the black silhouette of a child raising a fist of protest in front of a soldier, the background bright red.

I gave two quarters to the woman in the booth at the end of the bridge, then a bored federal police officer glanced at the water bottle and notebook in my tote bag and waved me on to Mexico. Avenida Benito Juárez was quiet in the late-morning heat, just a few men standing on its gray and black flagstones and smoking. The manager of a club called Faustos, with a Playboy bunny for a logo, had posted a handwritten sign on the door advertising that they were hiring pole dancers. Nearly every window nearby was blacked out, the doors bolted. On a few storefronts, the outlines of letters remained perceptible in the accumulated dust, residue from signs that had been peeled away after the businesses closed.

Avenida Juárez had been the bustling heart of the city up until it became a battleground in the late 2000s. This was midway through one of the most devastating periods of terror that any municipality has ever experienced, beginning a decade earlier when many of the single women flocking to the city from all over Mexico to find jobs in the maquiladoras—including some who were barely even teenagers—became the target of misogynist violence. More than 320 women were kidnapped and murdered in a ten-year period. Unbelievably, that was *before* Juárez became known as the most violent place on Earth. The first three months of 2008 saw more than 200 officially recorded murders. After the Mexican army was called in on what was dubbed the Joint Chi-

huahua Operation, the violence only got worse. By the end of the
year, more than sixteen hundred people had been killed. More
people were murdered in 2009, and then even more in 2010. By
the start of 2012, over ten thousand Juárenses had been slain in a
five-year period, thousands more than the number of American
soldiers killed in the wars in Iraq and Afghanistan.

A decade after the Sinaloa Cartel finished displacing the
homegrown Juárez Cartel and federal troops withdrew from the
city, scar tissue was obvious all over Juárez. Nevertheless, life in
the city continued much as it had before the narco war, including
for migrants. They had mostly avoided El Paso del Norte during the worst of the violence but began gravitating to the port
of entry again once things quieted down in the mid-2010s. At
Plaza de Armas, a number of vendors had set up stands in front
of the grand Juárez cathedral, including a cellular company with
a small tent and a team of young women in matching white tops
and green skirts dancing to "Hollaback Girl." Behind them, the
Santander had a long line of people waiting to use its ATM, while
two Haitian migrants rested on the curb.

Walking back toward the river, I found myself behind another
Haitian couple who were speaking in quiet Creole to each other
while, a few paces behind, their son absentmindedly swung a hat
to and fro. There were dentists and opticians on practically every
block—their practices primarily serving Americans who couldn't
afford to see a doctor in the United States—including in the
commercial spaces on the ground floor of the city's famous bull-
fighting ring. I asked a security guard where I could find the state
of Chihuahua's immigration agency, COESPO, and he directed
me to a building that abutted the Paso del Norte Bridge. In the
spare concrete landing below the metal staircase leading up to
the office, a number of families were gathered, speaking Brazilian
Portuguese, Turkish, and Mayan.

Upstairs, Enrique Valenzuela welcomed me into his immacu-
lately clean office, the only decoration a photograph of one of
the fifty-two border obelisks that were erected by surveyors after
the Treaty of Guadalupe Hidalgo and are engraved with the legend "Límite de la república mexicana." "Juárez has always been a

community used to people in transit," he told me. "A lot of people here have parents or grandparents who have migrated and gone to the U.S. or have relatives from earlier generations who still live in the south of Mexico. In 2018, we started to see an increasing number of people arriving in transit from the Caribbean and Central America. A lot of people from Cuba, some from Africa and eastern Europe. What did we do? The only thing we know to do when we see people in need."

Standing at a whiteboard in a blue-plaid shirt, the sleeves rolled up to his elbows, Valenzuela sketched out the ways that Mexican officials have responded to the United States' rapidly shifting border policies. Much of what he described reflected a reaction of migrants (and the polleros who guide them north) to the hardening of the border that began in the 1990s—given that it had become so perilous to attempt to evade the Border Patrol, in the 2010s the vast majority of migrants began simply walking into designated ports of entry and claiming asylum. International law requires refugees to be allowed to remain wherever they make their asylum claim until the case receives a legal hearing. Given that U.S. immigration courts were so overburdened that the wait time for those hearings was often numbered in the years, many refugees were eventually released within the United States.

In response to the overwhelming number of asylum claims they were receiving at the San Ysidro port of entry near San Diego in 2016, the Border Patrol there instituted a policy called "metering," whereby only a certain number of refugees would be processed on any given day. After assuming the presidency in 2017, Donald Trump expanded the practice across the entire border; by that fall, a significant backlog of asylum seekers had been created, all waiting in Mexico for their turn to meet with a Border Patrol officer. To streamline the process, first the Red Cross and then COESPO began maintaining a list of migrants in Juárez who were staying at a shelter called Casa de Migrante, waiting their turn in line. "It was mainly people from Cuba who wanted to be on the list, who recognized this was the system and wanted to go through the authorized port of entry," Valenzuela said, before rifling through a filing cabinet and showing me one of the metering lists from 2018. Each handwritten name was paired with a phone number and a country of origin, with El Salvador and Honduras competing with Cuba as the most frequent entry.

The situation became more complicated in January 2019, when the Trump administration adopted the Migrant Protection Protocols, better known as MPP, or the "Remain in Mexico" policy, which empowered the government to return migrants to the other side of the border while they waited for an asylum hearing instead of allowing them to stay in the United States. "We started to take in up to a hundred people daily," Valenzuela said. "That number went up to two hundred by June." In response, COESPO set up a provisional shelter at a high school gym, but since that facility only had capacity for about six hundred people, the agency was soon appealing to churches and NGOs for more support. Soon, a network of churches were hosting hundreds of migrants, with humanitarian agencies like the International Organization for Migration and UNICEF providing cots and bathrooms. "The Methodist church, they were used to having fifty people, tops. They got up to three hundred in the same space."

With the onset of the pandemic in March 2020, border entries were effectively frozen under Title 42 of the U.S. Code, which allows for the suspension of immigration to prevent the spread of communicable disease. "Things slowed down, people knew they couldn't get into the U.S.," Valenzuela told me. "But we can kind of anticipate what will happen with this phenomenon that is changing constantly." The result of the next big change—Trump losing reelection—was obvious. "Now President Biden said, 'No more MPP.'" Valenzuela snapped his fingers. Many of the migrants who had spent years waiting for their asylum hearing were allowed to enter the U.S., so even "those who had settled in Juárez said, 'I'm out of here, man.'" Then, in December 2021, the administration was forced to reinstate MPP because of a court order. Six months later, Valenzuela said, "not even fourteen hundred have been sent back, when back in 2019 we had fourteen hundred every week. It hasn't been that much of an emergency"—at least compared to the first go-around.

In an industrial section of East El Paso, ten miles away,

Ruben García settled into a metal folding chair in the kitchen of the labyrinthine warehouse that served as a migrant shelter for his organization, Annunciation House. Refrigerators and chest freezers were lined up against one wall, humming quietly. García apologized that the air conditioning was broken and offered me a bottle of water. As he told me his story, he paused now and then to gently mop his silver hairline with a wadded-up paper towel. "In 1978," García began, "I was part of a group that had spent about a year and a half discerning how to do something with our lives that was more meaningful, purposeful, that had some depth to it. That process eventually led to us deciding to place ourselves among the poor of El Paso and let them show us the way. At the time, there were only two homeless shelters, and neither one of them would accept undocumented people. So we made the decision to offer hospitality to them."

Annunciation House's original facility was the second floor of a small brick building on San Antonio Street owned by the Catholic Church. The group heard about an undocumented teenager who was living on the street and offered him shelter. Over the next forty years, the organization expanded in tandem with the growing number of migrants, and today its network includes eighteen facilities in Texas and New Mexico. "We're accommodating close to three thousand refugees every week, all of us together. They stay for short periods of time, the time it takes their sponsor to buy a plane ticket or a bus ticket and send them on their way."

García's phone rang. When he answered, the woman on the other end skipped a greeting, as if she were merely picking up a long-running conversation. "The first group is ready. I'm only going to have two groups today, because a lot of them did not qualify." The exchange lasted ten seconds. "That was the port of entry," García explained. "They're letting me know that they have Title 42 exceptions. If you're able to establish a fear of being in Juárez, the Border Patrol is supposed to let you in so you can continue the asylum process in the United States. Every day they've been receiving forty refugees—that doesn't mean they all qualify. Out of the forty there's usually two van loads, so that's maybe twenty-eight people." He nodded to the warehouse we were sitting in. "They're going to come here."

García gets these sorts of phone calls constantly, informing him of groups of migrants who have been processed by the Border Patrol and are ready to be released. He's almost always able to accommodate the new arrivals, though the government sometimes sidesteps the informal arrangement. A few weeks before we met, *El Paso Matters* reported that more than a hundred refugees had been dropped off downtown, the first such release in three years. "I'm not convinced that wasn't a political decision," García said. He had been informed of the release at five o'clock in the morning and was not asked to find space for the migrants. "They had four thousand people in their holding facility. If you really want to decompress the facility, you'd release one thousand. This was 119, and they were all single adults, heavily Brazilian, Turkish, Cubans—people that had the resources to take care of themselves. They were released at the bus station, formed little groups, went to the airport, and bought tickets on their own. They did not release families. People who were destitute. I think it was political, to send a message to the county and the city, this is what we can do if you don't create more space for us."

The tension between the Border Patrol and the city of El Paso came to a head a few months after our conversation, when Annunciation House and the smaller local shelters reached capacity because of an influx of refugees from Venezuela. Since the United States does not officially recognize Venezuela's government, it is legally obliged to allow those migrants to remain in the country while their claims are processed. In 2021, amid a contested election between the authoritarian President Nicolás Maduro and the reformer Juan Guaidó, Biden extended Temporary Protected Status designation to Venezuelans, and more than 150,000 of them arrived at the southern border over the next year. As their numbers in the El Paso shelter system ballooned in the fall of 2022, thousands were forced onto the street. In response, the state began chartering buses to send migrants to the sanctuary cities of New York, Philadelphia, Washington, and Chicago as part of an action dubbed Operation Lone Star. By the fall of 2023, some twenty thousand refugees had been sent to New York alone, and the Biden administration was forced to extend the Temporary Protected Status of Venezuelan migrants, in recognition that they would not be returning to their country anytime soon.

"You have people screaming, 'The border is out of control! We no longer have a country because we don't have a border!' So the answer is enforcement, enforcement, enforcement," García said. It's not that U.S. Customs and Border Protection, with its $17 billion budget, cannot afford to set up and maintain its own network of humanitarian shelters. It's that a political choice has been made to spend much more money on walls that force migrants into the desert, drones that surveil their movements, and officers who track them down, all in the name of a deterrence strategy that has seen the number of migrant apprehensions spike from 409,000 in 2016—when Donald Trump first ran for office—to 2.7 million in 2022. Nevertheless, anti-immigrant sentiment has only gotten more toxic, making even an avowedly religious organization like Annunciation House a target for conservative culture warriors. In early 2024, Texas Attorney General Ken Paxton accused García of "facilitating illegal entry to the United States, alien harboring, human smuggling, and operating a stash house" in a failed bid to shut his organization down.

"When I get up early, I like to go and sit on the porch with my cup of coffee and watch the caravans of people going down the freeway," García told me. "The truth is, millions of us relocate every year and we are *oblivious* to it. You have a son or a daughter, they just received their degree. They went to a university in Milwaukee, they get offered a job in San Francisco, they come and visit you in Albuquerque, then they move to California." In a typical year, a little under 10 percent of the American population moves: 30 million migrants. "It's very possible to have people arrive at the border and help them move in. From October 1st of 2018 to September 31st of 2019, in that fiscal year, immigration released 150,000 people to our network. You compare that to 30 *million* people? A hundred and fifty thousand is a drop in the bucket." Given that there are more than 330 million Americans, even the record number of migrants whom the Border Patrol detained in 2022 represents a fraction of a percent of the total population.

In lieu of any serious attempt to reform the existing immi-

gration system, conservative and liberal politicians alike have
coalesced around a policy of externalizing the border by forcing as many refugees as possible to remain in Mexico while their
claims are processed. (Indeed, in 2024 Biden issued an executive order barring many undocumented immigrants from seeking asylum—a clear echo of Trumpist border policy.) The dire
consequences of relying on a country significantly poorer than
the United States to manage this population of people in transit
became obvious six weeks before the expiration of Title 42, when
a fire broke out in a privately managed detention facility just a
block away from COESPO's office in Juárez. Surveillance footage
showed security guards fleeing the fire while dozens of migrants
remained locked behind bars. Forty people died in the flames.

The fire in Juárez only became international news because
the migrants died while they were in custody. Since the 2000s,
the migration networks that extend from Central America to the
southern border have been controlled by cartels, forcing those
fleeing from violence in El Salvador, Honduras, or Guatemala
to pay extra fees to the polleros who guide them north. If those
migrants make the mistake of telling their guide that they already
have family in the U.S., they are often kidnapped and held for
ransom. Women are frequently raped by their polleros or other
migrants. Even those who aren't attacked, held captive, or bribed
have to survive "La Bestia"—stowing away on the trains that connect southern Mexico to the U.S. border, usually by clambering
to the top of train cars and clinging to whatever handholds they
can find.

On La Bestia, losing your grip can be deadly. In his book
about the two years he spent accompanying Central American
migrants through Mexico, the Salvadoran journalist Óscar Martínez described a "man missing his right leg, a crutch under each
arm, stepping into the darkness toward the train tracks." The
man had lost his leg while attempting to ride the top of the train
near the southern city of Ixtepec, when he was inadvertently
pushed off by other migrants who panicked at the sight of lights
in the distance, which they mistook for police vehicles. He fell
between the cars, with one leg landing on the track. He was one
of the lucky ones: elsewhere in the book, Martínez describes a

child who was beheaded after falling on the tracks. Despite his near brush with death, after two years of recuperation the man was ready to try to board La Bestia again.

"Undocumented migration to the United States may fluctuate depending on the year," Martínez wrote, "but like a river it continues, ebbing and flowing, always finding its way to the sea." It doesn't matter how difficult U.S. authorities make the crossing. The conditions migrants face to just get to the border in the first place will already have forced them to draw on every ounce of resolve they have. That any of this is worth the danger is a testament to just how broken their countries of origin are. "They are not running from hunger—that most primal of needs—but from resignation," Martínez wrote. "They're unable to accept that miserable routine of waking up at five in the morning to travel two hours on a dangerous public transit system to . . . spend the whole day toiling away at undignified work . . . They are running, I believe, from the resignation that their children repeat the same miserable lives to which the depraved reality of Central America has condemned more than half of its inhabitants."

Since the turn of the twenty-first century, migration has become more global—though Mexicans, Salvadorans, Hondurans, and Guatemalans represented half of migrants arriving in the United States in 2023, the rest come from elsewhere around the world. I witnessed that shift firsthand in Juárez: Haitians fleeing the violence and instability caused by the 2021 earthquake and the subsequent assassination of President Jovenel Moïse, Turks fleeing the coup attempt against Recep Tayyip Erdoğan, Venezuelans fleeing an economy decimated by hyperinflation.

Embracing a more humanitarian approach to migration will only become more imperative as the climate crisis intensifies. According to modeling from *The New York Times* and ProPublica, as many as 30 million Central Americans may decamp for the United States in the next three decades as changing weather patterns further disrupt the region's agriculture. Paying Mexico to handle that influx indefinitely seems likely to only destabilize the country, perhaps triggering even more out-migration while exposing wave after wave of climate refugees to greater risk of banditry and sexual violence.

The physical and legal barriers to immigrants that authorities in the United States have erected since the 1980s function so poorly now that it's impossible to imagine them finding success in the future. Ultimately, we will have to face up to the moral necessity of accommodating a much larger refugee community than we do today. After all, the nation's imperial aspirations, so clearly articulated a century ago by William Smythe in *The Conquest of Arid America*, means that the United States has a hand in practically every crisis that forces innocent people to flee their homes. Reagan's interventionism in Central America begat Bush's wars in the Middle East begat Biden championing regime change in Venezuela. The longer we wait to accept the responsibility that comes with wealth and power, the more painful the process of contending with the human cost of our globe-spanning empire will become.

After I finished speaking with Enrique Valenzuela in Juárez, I crossed over Avenida Rafael Pérez Serna and stood on a dirt landing above the concrete trench that a map would tell you is called the Rio Grande. The engineers who canalized the river in the 1920s called their work the Rio Grande Rectification Project, as if the sixty-seven miles of snaking, riverine habitat that had been removed was a mistake in need of fixing. After the Rectification Project was completed, the commissioner of the International Boundary Commission observed that, in addition to avoiding any future frontier disputes, the concrete channel would create "more satisfactory enforcement of the immigration and customs laws of both" the U.S. and Mexico.

The American wall stood on the opposite bank of the Rio Grande, the office buildings of downtown El Paso visible just above its crown of concertina wire. I tried to reconcile the sight with the river I have known and loved my entire life: the easy curves, the shady cottonwoods, the soft flow of gritty water. I remembered peeling off my shoes and socks to wade in the river on a particularly hot day of wandering the Bosque in Albuquerque. Standing on the bridge over the Rio Grande Gorge near Taos, trying to understand how such a thin ribbon of current could cut so deeply into the landscape.

Here, in the exact crux of El Paso del Norte, the Rio Grande had been reshaped by politicians and engineers into a matter of concrete and politics, a sad, fearful barrier rather than an ecology. Gloria Anzaldúa wrote that the border is "una herida abierta where the Third World grates against the first and bleeds." But she drew a distinction between the frontier itself and the landscape it had created, writing, "A border is a dividing line, a narrow strip along a steep edge. A borderland is a vague and undetermined place created by the emotional residue of an unnatural boundary. It is in a constant state of transition. The prohibited and forbidden are its inhabitants."

Today, more than 40 percent of the United States is nonwhite, with Anglos expected to become a national minority in the 2040s. That notion feels theoretical to many Americans east of the Mississippi, who still largely inhabit cities and suburbs that look much as they did a half century ago. But in the Southwest, the majority-minority future is already here. Albuquerque, Phoenix, Tucson, Las Vegas—none of these cities is more than 43 percent white. El Paso, meanwhile, is 80 percent Hispanic, the highest proportion of any big city in the country. And of all the states these cities occupy, only in Arizona do Anglos still—barely—outnumber all other ethnic groups.

The profound demographic shift unfolding across the United States is only partially being driven by immigration from Latin America and Asia, the points of origin for roughly 10 percent of the population (one out of four of whom are undocumented). Just as important is the rise of multiracial Americans, who almost tripled in number between 2010 and 2020, and now represent another 10 percent of the country. Part of that rise has to do with the demographic of people who are partially Asian American, whose growth has outpaced that of those who identify as Asian American alone. At the same time, the changing self-image of many Hispanics has contributed to a boom in multicultural identification: while 50 percent fewer Hispanics described themselves as Caucasian in 2020, the number claiming "two or more races" rose by 567 percent from the previous census.

While America is only becoming more multicultural, the

legacy of urban renewal and segregationist housing policy means the pressures of climate change are falling disproportionately on nonwhite neighborhoods, where water access is less secure and heat waves more intense. Creating a more equitable country will require closing those persistent gaps, work that's already well underway in the immigrant communities that are forcing powerful interests to attend to their needs through collective action.

When Vida Lin first moved to Las Vegas from California in 1994, she found the city's Asian communities to be fragmented. "I probably attended twenty different Chinese or Taiwanese groups. There was the Japanese American Citizens League, the Thai associations, the Korean groups," she told me. "They were all little silos. Most of them taught ESL classes, and the big churches and temples would do cultural celebrations. But when it came to issues, they kept quiet. 'We're the foreigners, we want a better life for our family, let's keep our head low.'"

However balkanized, Las Vegas's Asian and Pacific Islander communities were growing quickly, surging from 3 percent of Clark County's population in 1990 to 12 percent thirty years later. Much of that had to do with the efforts of early immigrants like the singer Sue Kim. After Kim became an American citizen, she sponsored the emigration of more than forty family members from Korea to Las Vegas, often helping them find homes once they arrived, as well as well-paying casino jobs. Once she retired from performing, Kim got a real estate license and specialized in working with the immigrants who were arriving en masse in the '90s. That decade saw the establishment of Vegas's Chinatown, which started out as a single strip mall. Now, Chinatown stretches for three miles along Spring Mountain Road.

It may have once been possible for a handful of early immigrants and business owners to play an outsize role in getting new arrivals settled, but with more than 250,000 Asians and Pacific Islanders in Clark County, Vida Lin became convinced that the community would need more formal advocacy if it was to prosper. "People in the United States would always try to divide different groups from each other, right? But when we come together, we can have power."

In 2015, Lin founded the Asian Community Development Council of Nevada. Initially, its efforts mostly revolved around registering tens of thousands of new voters for the next year's presidential election under the theory that "if we get our community to the polls, we'll have a voice." Gradually, the organization evolved into a much broader community resource, including supporting service workers who lost their job during the pandemic. "We opened up a culturally sensitive food bank. We worked with Cisco to do this big food donation, which was great, but they were giving us cheese and frozen chicken nuggets—we don't know what to do with that! We did a food drive where we had rice and fish sauce, vegetables from our community. That program, we gave out 2 million pounds of food." After the health crisis became less acute, Lin's organization shifted its focus to education, working with city officials to bring Asian studies courses into public schools.

"Being in Las Vegas for over twenty-eight years, we are one of those cities that can reinvent ourselves any time," Lin said. "We mutate, we change. It used to be gambling only, then it became about conventions. Then it was an entertainment capital. And now we're in sports. We have the Raiders, the Golden Knights. It's crazy! There's no city like that in the last twenty, thirty years." Lin wants to make sure Asian Americans continue to share in Las Vegas's dynamism, whether they have roots in the city or only moved there last week. "There's no enemy here, let's not divide people. Let's understand the city is home to lots of different cultures and embrace it."

In a region with as deep a history as the Southwest, continuing to celebrate the heritage of the longest-tenured residents is just as important as welcoming new arrivals. To do both, there's no more powerful tool than the primordial language of food. Few understand that better than Don Guerra, who owns a Tucson bakery called Barrio Bread. When I met Guerra outside the shop, he was wearing a loose-fitting shirt and aviators, with a silver medallion around his neck that glittered in the sun. "I come from a food family," Guerra said. "My mom used to make all of our bread—we didn't have a lot of money. But I loved it." He chuckled about trying to snatch freshly baked bread from the counter, remembered watching hungrily as his grandmother rolled out fresh tortillas.

After we sat down in the back of the restaurant next door, Guerra told me he had grown up in Tempe and learned his first lessons in entrepreneurship at his father's barbershop, where he was responsible for shining shoes starting at the age of eight. His dad told him, "You work for tips, and then I'll give you ten bucks at the end of the day. But save your money because you're going to have to reup your supplies. That comes out of your cash, not mine." After he turned thirteen, Guerra got a job in a local Greek restaurant, working his way up from bussing tables and washing dishes to cooking on the line. By his senior year of high school, he was running the kitchen.

Obsessed with mountain biking and snowboarding, Guerra moved to the modish old railroad town of Flagstaff after he graduated from the U. of A. (Flagstaff is nestled in the foothills of the magisterial San Francisco Peaks, only a short drive from the Grand Canyon.) "My first day in Flagstaff, I met someone who worked at a bakery. They said come tonight and see what we do and if you like it. It was a three-person operation, we'd make bread from four in the afternoon until midnight for the next day's service. I went from restaurants where it's hustle, hustle, hustle— the tickets come, you've got to pump food out—to bread, where it's like: Everyone calm down, we've got to bring good energy to the bread. You have to be patient, mindful. You have to nurture this product along."

After studying under a French pastry chef at the Boulders Resort in Carefree, Guerra opened the Village Baker of Flagstaff in 1995, which he expanded to Ashland, Oregon, two years later, thinking he'd be able to do some serious biking in the Pacific Northwest during his downtime. But hurtling back and forth over the nine hundred miles that separated the Village Baker's two locations proved to be unsustainable. "I eventually blew up because I got tired and frustrated. I was too young to manage all the people, I had like thirty employees. They wanted to be my buddy, and I was trying to be a nice guy—they just ran me ragged." Guerra sold his stake in the company and spent six

months trying to reset his life through physical exertion, training for triathlons and packing peaches with migrant workers in
southern Oregon. When that didn't work, Guerra decided to
return to Tucson, where he settled into teaching physical education and math in the city's public school system.

Only after he'd spent seven years as a teacher did Guerra
finally feel an inclination to try to get back into food. Never
one to do anything halfway, Guerra converted his garage into
a bakery and started selling bread at farmers markets, modeling
his business off the grassroots home bakeries that are common
in Mexico. But before he was ready to leave teaching behind,
Guerra realized that he needed to take a different approach than
he had at the Village Baker. "I'm like, I need to find a way to do
it differently. I don't want to be the same baker using commodity
flours."

Guerra started making bread for a local nonprofit called
Native Seed Search, which conserves the endemic and heritage
crops best adapted to growing in the Southwest. They had just
won grant money to study and cultivate chapalote corn, the variety that was first grown in the Tucson Basin four thousand years
ago, as well as white Sonora wheat, which was originally planted
by Eusebio Francisco Kino, the Franciscan friar who founded the
first mission in the Tucson Basin, San Xavier del Bac, in 1692.
"The first night I baked bread with the flour they had grown,
it was like the first time I had made bread in Flagstaff. I had a
vision: Okay, I'm a baker, but now I'm going to do it with local
grains. I bought a little canister mill—it's called a Wonder Mill—
used that to mill the seeds, make a flour. Now that I have flour, I
can make bread. It didn't turn out great, it was very rudimentary
bread. But it was a start."

Since white Sonora is a soft wheat that is better suited to
making tortillas than leavened bread, Guerra suggested that the
local growers who partnered with Native Seed Search should also
plant durum and hard red wheat, which could then be blended
with the heritage varieties. To make it worth their while, Guerra
promised to buy whatever they produced. From starting out having to use 90 percent commodity wheat with only 10 percent
locally grown grains, he slowly developed enough of a relation-

ship with farmers to flip those proportions. "I'm pretty close to sustainable, using only farms within a hundred-mile radius. This has been my ultimate mission and vision for myself as a baker: Barrio Bread as a model of a community-supported bakery working to develop a local grain economy." His agricultural collaborators include Brian Wong, whose family has operated a farm in Marana since the 1930s, and the co-op the Tohono O'odham Nation operates on the San Xavier Reservation.

Though Guerra has become a culinary celebrity in Tucson, he insisted, "I'm not so worried about building an empire—I'm more into, how do I serve my community? How do I do it in a way where I have the most impact for them? It's not just, 'trade me money for bread.' I'll teach you how to do it too if you're interested, so people have less reliance on me." During the pandemic, Guerra said he'd given away his sourdough starter for free and offered online baking classes, helping his customers learn the craft of baking for themselves while they were quarantined. "That's my educator background. Share the knowledge, help them create some excitement and sustainability, food security, in their own lives."

Whether through tailoring a food bank to Asian immigrant populations or fostering a market for heritage grain, food, in the Southwest, represents a point of connection between the many cultures that call the region home. Even dishes that are thought of as traditional in New Mexico and southern Arizona, like calabacitas—a mixture of corn, squash, and chile—represent a fusion between autochthonous crops and vegetables that were brought north from Latin America. Over the centuries, the foodways of Puebloans and Mexicans melded into something singular and lasting. With time, the same may happen all over America as new immigrants embed themselves within the country's fabric, so long as those whose ancestors arrived less recently accept that the nation has always been at its best when it has celebrated its mutability.

La Jornada and *Numbe Whageh* are located just a few blocks east of Old Town Plaza in Albuquerque. The pair of artworks speak to New Mexico's enduring struggle to tell its story in a way that's equally agreeable to all the people who call the state home, an uneasy compromise that was reached after years of civic debate—and then revised.

The erection of the twin monuments was set in motion in 1996, when a woman named Millie Santillanes proposed commemorating the four-hundredth anniversary of the conquistador Juan de Oñate's establishment of the permanent colony of New Mexico in 1598. Santillanes was a lifelong resident of Old Town who could supposedly trace her genealogy back to one of the Spanish families that first settled Alburquerque in 1706. In short order, a plan took shape to commission a bust of Oñate that would be displayed on Old Town Plaza and titled *Cuarto Centenario*.

When Albuquerque's Indian community heard that the city's arts board had approved a project that would give Oñate central billing in Old Town, there was an uproar. In response, the board amended the project in 1997 to mandate that an Indian artist be appointed to work with the artists who had already been hired for *Cuarto Centenario*, the Anglo sculptor Betty Sabo and Reynaldo "Sonny" Rivera, best known for having completed a massive statue of Oñate in 1993 at a visitor's center in the town of Alcalde, about halfway between Santa Fe and Taos. The board chose Nora Naranjo Morse, a Tewa ceramicist and landscape artist who is a member of the Santa Clara Pueblo, for the job, completing a trio that supposedly gave equal representation to each of the state's three cultures.

The team immediately descended into dysfunction. Core to the conflict was the artists' differing, deeply held views about the history they were supposed to be commemorating. Rivera and Sabo saw Oñate as a heroic founding father, while Naranjo Morse believed he was a tyrant. Complicating matters further was a lack of cohesion between the artists' styles: both Rivera and Sabo were figurative artists, while Naranjo Morse was not. Eventually, Rivera and Sabo stopped speaking to Naranjo Morse entirely. "The city was probably looking for someone who would be a token," Naranjo Morse told the *Alibi*, the city's alternative newspaper. "I was supposed to collaborate quietly." Surely it didn't help matters when, in late 1997, Rivera's statue of Oñate in Alcalde was defaced: under the cover of night, Indigenous activists amputated the right foot of the bronze figure—symbolic revenge for the conquistador's sixteenth-century order to sever the same right foot from hundreds of Acoma men to punish them for opposing Spanish rule.

In 2000, Albuquerque's city council stepped in, deciding that *Cuarto Centenario* should now have two components, one Native American and one Spanish, the latter emphasizing anonymous settlers rather than Oñate himself. Sabo sided with Rivera and contributed to the Spanish portion of the artwork. The compromise did not, however, include equal compensation for the three artists. According to the *Alibi*, the team of Rivera and Sabo were collectively paid $150,000 for the project, while Naranjo Morse got only $22,000.

The result was unveiled in 2004, six years after the actual colonial quadricentennial. The Spanish half, *La Jornada*, consisted of a wagon train of bronze statues, with Oñate out in front of a party of Spaniards that included a Franciscan friar, playing children, and two figures whom Sabo modeled on Santillanes and Mayor Marty Chávez. A few yards away, on the other side of a moat of bare dirt, the Indigenous half of the installation, *Numbe Whageh*, spiraled gradually into the ground. The first time I visited the artwork, it looked, from afar, like a small hill covered in chamisa, scrubby juniper trees, and mountain mahogany. But as I followed a path inside, I descended deeper into the earth than I could have previously expected; it was a blazing summer day, and the feeling of reprieve *Numbe Whageh* granted was reminiscent of being engulfed by shade trees. At the end of the path, enclosed by high banks of earth, was a small spring so well insulated from the noise of the nearby streets that I could hear it burble.

Numbe Whageh is a Tewa phrase that roughly translates to "center place," with the sense of a center being meant both physically and spiritually. Reemerging from the artwork was startling, as the curves of the sanctuary were replaced by the exacting lines of the concrete sidewalk and the traffic churning between Mountain Road and 19th Street. Naranjo Morse told an interviewer from the radio show *Native America Calling* that this "visual contradiction" was intentional. "It says something about who we are as Native people holding on to the idea of what is important, what our ancestors said was important. The land."

Initially, the most generous reading of the pairing of *La Jornada* with *Numbe Whageh* was that experiencing the sculptures together forced a conversation between Spanish and Indigenous cultures. To my eye, the defiant features of the bronze sculptures read instead as a charged refutation of the reflective space created by *Numbe Whageh*, while the landscape sculpture's sense of apartness had a bittersweet quality. Visiting the spot was, in some sense, a vision of New Mexico at its worst: the stark beauty of the desert rendered banal by heedless conquerors intent on imposing the same European lifestyle that they previously had on the Caribbean, Mexico, and Peru.

Activist resistance to *La Jornada* dimmed after it was finally unveiled, but was never entirely extinguished. In the summer of 2020, calls for a new historical treatment of Oñate were renewed after George Floyd was murdered by a police officer in Minneapolis, setting off a wave of worldwide protests. As the demonstrations grew in scope to call for societal changes that went beyond merely reexamining the role of law enforcement, statues of other men who had committed atrocities against nonwhite communities became targets of anti-racist rage. Shortly after a statue of Columbus was beheaded in Boston and the figure of the British enslaver Edward Colston was heaved into the harbor of Bristol, protestors in Albuquerque demanded the removal of Oñate from *La Jornada*. On June 15 that year, a group protesting in front of the statues was met by armed members of a white supremacist militia; no police officers were present, and the militia eventually began firing at the crowd, injuring one of the demonstrators.

Oñate's statue was removed the next day, though his retinue remained in place. A week later, Sonny Rivera penned an op-ed in the *Albuquerque Journal* defending his work. After arguing that the Acoma had provoked Oñate's violent retribution and describing how Spanish treatment of Indians in New Mexico had been far better than what eastern tribes faced from the English ("today New Mexico Native Americans still have their own lands and their language and their religion"), he returned to the original

tri-cultural triumvirate compromise that facilitated the placement of the two sculptures together, implying that the contro-
versy about Oñate's inclusion had already been resolved. "Maybe
some want to tear open old wounds between Native Americans
and Hispanos," Rivera wrote. "The wound was not caused by my
work; it was already there."

Deb Haaland, then representing Albuquerque in Congress,
struck a more conciliatory tone in an editorial that ran the same
day as Rivera's. "I know our history is painful," she wrote, "but
we are all New Mexicans." Haaland is a member of the Laguna
Pueblo and has lived in Albuquerque since the 1970s. For many,
her life story demonstrates the ability of Indians to overcome the
discrimination they habitually confront in urban centers: from
a homeless single mother putting herself through law school at
UNM by selling salsa out of her car, Haaland rose to become the
chair of the state Democratic Party, a two-term congresswoman,
and then the first-ever Native American to serve as a cabinet secretary, when she was appointed to head the Department of the
Interior in 2021.

In the call for unity Haaland authored in the *Journal*, she did
not excuse Oñate, even pointing out that the Franciscan priests he
brought to New Mexico "enslaved my ancestors." Her goal was
to refocus the conversation to the ways in which Pueblo and Hispanic culture have grown together in the intervening centuries,
noting their shared foodways and struggles against white oppres-
sion. "Both Hispanic and Native Americans have experienced
systematic racism," she wrote. "Native children were carted off
to boarding school, separated from their families. Generations of
Hispanic students were punished for speaking Spanish in school.
But we haven't allowed those injustices to derail our drive to
move our state forward." She concluded: "This is our state, and
we must celebrate who we are: New Mexicans."

Haaland's op-ed was a valiant effort, but America, unfortunately, still abhors any call to embrace our contradictions. Around
the same time Oñate was removed from *La Jornada*, Rivera's
original statue of the conquistador in Alcalde was taken down
by the manager of Rio Arriba County. In 2023, the local govern-

ment decided to bring the statue out of retirement: this time, instead of being placed at an out-of-the-way visitor center, Oñate would preside in front of the county's main offices in Española, a small city about twenty-five miles north of Santa Fe. In protest, the Red Nation, an Albuquerque-based activist group, staged a rally in front of the county offices in late September.

Jennifer Marley, from the Pueblo of San Ildefonso, was one of the organizers, having previously participated in the protests that led to the cancellation of the Entrada pageant in Santa Fe. Standing in front of a banner that read "1680: Celebrate Resistance Not Conquest," Marley told supporters, "Oñate represents the death drive of colonialism." Despite her more heated rhetoric, Marley's message paralleled what Haaland had written a few years earlier. "Let's clear something up. This is an issue that is falsely framed along racial lines. This is not Natives versus Hispanos, that's fucking ridiculous. This is an issue of colonialism versus decolonization . . . All of our ancestors were victims of U.S. settler colonialism. And what's wild is that, when we're focusing on all the clashes, we're totally distracting from all of the history where we were in solidarity together. Native people and Nuevomexicanos and Genízaros."

Rio Arriba County backed down, announcing that the planned installation of the Oñate statue would be postponed. To celebrate, the Red Nation organized a prayer vigil on the site. It was a typical fall morning in Española—sunny but cool. A few dozen activists showed up, draping the concrete plinth where the statue was supposed to be installed with tall stalks of corn, handwritten signs reading "Land Back," and a black hoodie with an image of an Indian woman screen-printed on it, "JUSTICE" stamped over her eyes in red.

As the activists stood and prayed, a few of them holding banners, another group gathered off to the side. One of them, a twenty-three-year-old with long brown hair spilling out of his red "Make America Great Again" hat, walked around shaking hands with the other counterprotestors and telling them, "Thank you for standing with the rule of law." Then he went with a few others to take pictures of the prayer circle. One of the demonstrators, who had a Zia symbol tattooed next to his eye, approached the men, telling them, "You guys aren't going to take a picture, cause you're doing it for the wrong reasons." The guy in the MAGA hat snarled, "Watch me."

More activists blocked the man from approaching the vigil, telling him to cool down. They managed to push him back toward the parking lot, before he pulled a gun from his sweatshirt and charged. The activists swarmed him, pushing him back again while someone, perhaps not realizing he was armed but alarmed at the violence, yelled, "Hey, let him go, let him go!" Once they did, he stood back and aimed, shouting, "Fuck you, bitch!" as he fired.

A Hopi and Akimel O'odham activist named Jacob Johns collapsed to the ground, shot in the gut; the man swung his pistol toward another demonstrator and appeared to try to fire again. This time the gun jammed. He ran back to his silver Tesla and fled the scene, while organizers rushed to help Johns. Prior to the prayer vigil, Johns had been preparing to lead a delegation of Indigenous Wisdom Keepers to an upcoming international climate conference in Dubai. Now he was on the pavement, blood pulsing from his stomach. Only after being airlifted to a hospital in Albuquerque and undergoing two surgeries did Johns's condition stabilize. Meanwhile, the shooter was apprehended by police from the Pueblo of Pojoaque. Hours later, county officials announced that they would not move forward with a reinstallation of the Oñate statue.

The confrontations in Albuquerque and Española serve as chilling reminders that too many Americans still reject an honest accounting of the nation's history and are willing to resort to violence to preserve white supremacy. Even the lines by Emma Lazarus that are meant to give voice to the Statue of Liberty are misleading:

> *Give me your tired, your poor,*
> *Your huddled masses yearning to breathe free,*
> *The wretched refuse of your teeming shores.*
> *Send these, the homeless, tempest-tost to me,*
> *I lift my lamp beside the golden door!*

For all the stirring sentiment of "The New Colossus," the European immigrants who were processed at ports of entry like Ellis Island around the turn of the twentieth century were not made especially welcome. This was the period when most of my ancestors first arrived in this country—Neapolitans and Sicilians to Boston, eastern European Jews to New York and Chicago. They ended up in crowded tenements, struggling for a foothold in their adopted home. Luckily, these ancestors made it across the Atlantic before 1924, when the nation fully abandoned its experiment with open immigration by passing the Jackson-Reed Act. That legislation set annual quotas for countries of origin pegged to the 1890 census—when there were not yet significant southern or eastern European communities in America—and extended the forty-year-old prohibition against Chinese immigrants to almost all of Asia. It was not until 1965 that the quota system was lifted and Asians began to be able to lawfully emigrate again.

The unity offered by monuments like the Statue of Liberty is illusory. Multiculturalism requires all of us to participate, regardless of race, ethnicity, or immigration status. It is a messy idea, seeded with the same sorts of conflicts that New Mexicans have been wrestling with for four hundred years. So be it. Being American has never been simple.

Most years, my dad and I try to spend a day or two fishing the San Juan River, just south of the Navajo Dam where the cold, clear water released from the earthwork embankment creates an ideal habitat for trout. We always stayed at Abe's until it closed, then moved across the street to a place called Fisheads. Now that neither of us lives in New Mexico and it's a pain to fly with waders, my dad usually springs for a float trip: a guide picks us up in his truck, we tow a ClackaCraft to the put-in spot right below the dam, then we spend the day drifting downriver. Unless it's an unusually quiet weekend, we'll ask to jet through the crowds at Texas Hole and spend a bit more time in the stretch past Simon Canyon, where there are fewer boats, the riverbanks are thick with reeds and dragonflies, and we can always get into some fish among the nested rapids.

Looking upriver, the Navajo Dam might as well be another wall of the canyon—the only real differentiation is that, instead of tiers of sandstone studded with piñon, it's a gradual dirt slope coated in dry, yellow grass. Only once you're on top of the dam do you start to fathom all the water that constitutes the San Juan: the river drains the western slope of the San Juan Mountains in southern Colorado, with the eastern slope feeding the Rio Grande. By the time the San Juan has merged with the Piedra, the Animas, and the rest of its tributaries, the river is carrying melted snow that first settled more than fourteen thousand feet above sea level, at the peaks of mountains named Eolus and Wilson. What isn't impounded by the Navajo Dam gets caught by the concrete edifice in Arizona's Glen Canyon, which has refigured the confluence of the San Juan and the Colorado into the reservoir of Lake Powell.

The Glen Canyon Dam cuts a more alarming profile than its rammed earth counterpart, 240 miles upriver. A true feat of engineering, the Glen Canyon Dam is formed by a concave arc of concrete bolted deep into the crimson rock at its sides, the crest rising more than seven hundred feet above the Colorado. There's no fishing down there: the walls of the canyon are simply too steep. On the bluff above Glen Canyon, a whole mess of transmission towers preside over the switchyard where electricity from the eight generators inside the dam is converted and distributed all across the service territory of the Western Area Power Administration, from Nevada to Nebraska.

After grokking the dam, I drove to the dusty lot a few hundred yards away from its lip. On foot, I followed signs for a trail that led north, up onto the slickrock where I could admire the canyon country's stupendous buttes and plateaus rising in jagged layers out to the blue hump of Navajo Mountain, thirty-five miles distant. The water of Lake Powell had a deep sapphire tint. The path jogged right and ascended again. Around the next curve, I found myself confronted with a protruding shelf of pink rock. Under its shadow, the wall went entirely green: a vertical carpet of maidenhair ferns. As I stepped under the overhang, the air temperature dropped 20 degrees. I sat for a while to cool off, resting and enjoying the hospitable smell of dirt and water.

SPRAWL

The deeds and papers don't mean anything. It is the people who belong to the mountain.

—LESLIE MARMON SILKO

In the southeast corner of the Las Vegas Valley, the Cowabunga Bay Water Park's corkscrewing water slides and sixty-foot-tall plastic banana—home to "the world's largest manmade wave"— sit kitty-corner across the eight lanes of I-515 from the Galleria mall and the Sunset Station Casino. Just down the road, a sign in a vacant lot surrounded by a chain link fence announces that ranch homes are now leasing at a new development called Marble Mesa. After I drive past a concrete plant, a utility lot, and one of the city of Henderson's water treatment facilities, the industrial trappings of exurbia drop away to reveal an open expanse populated by bulldozers pushing dirt into piles, as if building a megalithic ant colony. Another sign suggests that this is the future site of Glenmore, a "single story community."

Glenmore is but one of the numerous neighborhoods in Cadence, a development encompassing some 2,200 acres of a former dump that will eventually include homes for more than thirty thousand people. That makes it only the latest in the long line of master-planned subdivision agglomerations that constitute Henderson. Originally described by a government brochure as "an emergency-born community development" to house employees of a magnesium-processing facility during World War II, Henderson is now the second-largest city in Nevada, with a population of more than 330,000.

I drive through a roundabout lined with orange construction barrels and take the exit heading north. The right side of the street is occupied by a community called Weston Hills, which encircles the Chimera Golf Club; to the left, a fence lined with barbed wire marks off a field of gravel. At the end of the road, I find a red-dirt parking lot that facilitates access to the Las Vegas Wash, the channel that funnels the metropolitan area's recycled water back into Lake Mead. Four pickup trucks are parked there, though the picnic area's concrete plaza is vacant beneath its maroon sunshades.

I get out of the car and head to the Pabco Weir by hopping a concrete barricade meant to steer people away from the wash itself and to the walking and biking paths that run alongside it. Despite the posted sign stating that fishing is prohibited on the weir, a dude in Air Jordan basketball shorts and matching slides is doing just that: squatting on the concrete embankment and casting a line just under the small waterfall created by the weir, one of the twenty-one structures that control the rate of the channel's twelve-mile run through the valley. On the other side of the wash, a crew of four teenagers in all black loaf in the sun. One of them has a dirt bike that he keeps trying to ride up the concrete slope of the weir and then kick up into a tail tap, but he never quite nails it.

Overhead, three pelicans with yellow beaks and black-tipped wings swoop into view, wheeling into an updraft while, in the elbow of the weir, a few cans of Monster Energy drink bobble in the foam. I bushwhack downstream through the cottonwoods that line the wash until I find clear passage to the water by way of a massive desert tobacco plant, dense with white, flute-shaped flowers.

While the Las Vegas Wash originally functioned as a natural drainage system for the valley's stormwater, it now serves as the channel through which 200 million gallons of water are returned to Lake Mead from the city every day. It closes the loop of the cycle that begins on Saddle Island, just five miles north of the Hoover Dam, where three intakes extract water from the reservoir and ferry it to the Southern Nevada Water Authority, which depends on Lake Mead for 95 percent of its supply. After the authority serves the 2.3 million people who live in Clark County, sewage is collected in massive recycling facilities, treated in a three-stage process, then, once clean, funneled back into the wash.

Contrary to the reputation for profligacy generated by the Bellagio's fountains, the copious golf clubs, and attractions like Cowabunga Bay, Las Vegas is among the most efficient municipal water users in the world. While all of those public accommodations are indeed massively wasteful, they represent only 60 percent of the city's consumption. The rest is used indoors, and virtually all of it is recycled. In addition, Las Vegas's emphasis on outdoor conservation cut per capita water use in half between 2002 and 2022, a period that saw Clark County add more than seven hundred thousand residents. The region's remarkable efficiency means that, even after Nevada's allocation of water from the Colorado River was cut by 8 percent in 2023, it still used tens of thousands of acre-feet less than it was entitled to.

Much of the city's contemporary relationship to water can be attributed to the efforts of the pugnacious Patricia Mulroy. After growing up on an Air Force base in Germany, Mulroy moved to Las Vegas in 1974 when UNLV offered her a scholarship to study German literature. On her first night in town, she slept in the red velvet bedding of the Desert Rose Motel, directly beneath a mirror attached to the ceiling. Though she'd initially aspired to work as a diplomat overseas, Mulroy was scared off that path by the pervasive sexism of the State Department, which prevented most women from rising to any position of authority. Instead, she found her way into a much more intimate level of government by working as an analyst for Clark County, where she helped local officials lobby for state funding in Carson City. In 1985, she was promoted into a leadership post at Las Vegas's water authority.

Though Mulroy had no experience with water policy, she was a savant when it came to amassing political power. Quickly, she realized that the infighting of the assorted utilities in southern Nevada was preventing them from addressing a trend that threatened them all: a vertiginous rise in water consumption. "We were seeing 17 percent annual increases of water use in the late '80s," Mulroy told me. "There was no conservation. We were planting Kentucky bluegrass and East Coast trees, doing everything we possibly could to replicate a more lush environment and detract from the stone desert that we live in." Though Clark County had recently completed its first intake from Lake Mead, allowing it to enjoy Nevada's entitlement to three hundred thousand acre-feet of the Colorado River, spiking consumption meant that the individual municipalities of Las Vegas, Henderson, Boulder City, and North Las Vegas began to see each other as rivals in a zero-sum game of securing water for new development.

After Mulroy won the top job at the Las Vegas utility in 1989, her first initiative was an audit of the letters that had been issued to builders guaranteeing them the water they needed to secure financing for new subdivisions and resorts. "What it showed was that we were way overcommitted," Mulroy said. "There was *no way* we could meet all those commitments." She ordered a temporary moratorium on new development, which spooked her peers in North Las Vegas and Henderson into taking a look at their own books, where they found similar issues. "It became pretty obvious that there was only one way out, and that was for us to combine forces."

All the previously competing interests were corralled into the Southern Nevada Water Authority, an entity that could both manage the valley's existing resources and seek to supplement them. Almost as soon as the SNWA formed with Mulroy as its head in 1991, a report was issued stating that Nevada would surpass its budget of Colorado River water by 1995. Though Mulroy contended that the region could actually hold out until 2006, the state mobilized around bolstering its water resources so that growth wouldn't have to be curtailed. Steve Wynn was quoted in *The New York Times* about his fears that water restrictions would "put an imaginative damper on the future"; a few years later, Mulroy herself told *High County News* that "the bankers were starting to get squeamish about giving out loans . . . It was looking like it was going to undermine the whole economic fiber of southern Nevada."

Scaling up water recycling was the least controversial way to addressing the shortfall. "Once we were working together," Mulroy told me, "we could enter into a contract with the federal government to access return flow credits. The reason we're so good at recycling is because, for every gallon of wastewater that we treat, put in the Las Vegas Wash, and return to Lake Mead, we can take a like amount out. If we're using two hundred thousand acre-feet and returning 180,000 acre-feet—do the math." In 1992, Mulroy lifted her moratorium and the city's growth was fully unleashed, helped along by the fact that the '90s ended up being one of the wettest decades ever recorded in the Colorado River Basin. All the excess moisture meant that, even after accounting for its return flow credits, southern Nevada was able to simply consume more water than it was entitled to.

Sin City was still seen by most Americans as little more than a tourist trap with a libidinous edge when Mulroy first moved there in the 1970s. By 1999, casino management had fully transitioned from organized crime to publicly traded companies and Vegas was the fastest-growing city in the nation. While, as recently as 1980, Las Vegas, Albuquerque, Tucson, and El Paso had all been roughly the same size, the half-million people who moved to Clark County over the course of the '90s far surpassed the combined number of new residents the other three managed to attract. By the turn of the twenty-first century, Las Vegas was the second city of the Southwest, the only one that could possibly rival Phoenix when it came to economic impact and national swagger.

The flip side of Las Vegas's efflorescence was that the city was less nimble when the dry times returned. In 2002, the Colorado only carried a quarter of its previous average flow and the stockpile in the two reservoirs shrank by 18 percent. Mulroy referred to that year as "cataclysmic." In response, she instituted new rules governing how water could be used outdoors. Unlike the Phoenix Water Service, which had been able to draw down water consumption solely by raising the cost of water in the summer, "We began a program that was carrot and stick. The carrot was, we will pay you—we started at a dollar and then quickly went to two dollars—per square foot to take your grass out. We put the golf courses on a water budget. We began to fine people for over-irrigating. And we tried some things that were less successful. When we tried to ban washing your car in your driveway, I had every senior citizen in southern Nevada after my throat. I had no idea that in a senior's routine, washing his car is one of *the* events of the week!" Nevertheless, after only two years, the reforms had brought the district's annual consumption of the Colorado down to 273,000 acre-feet—well within Nevada's allocation.

Mulroy supplemented her conservation campaign by bolstering Las Vegas's water recycling capabilities with the River Mountains Water Treatment Facility, which can process 300 million gallons of water every day. She also pushed to create a new intake into Lake Mead, a "third straw" that would allow the utility to pump from the bottom of the reservoir, justified as a means of improving the quality of the county's water (the deeper an intake is submerged, the lesser the need to filter out particulate that descends into the reservoir from its surface). The new intake was well underway in 2014, when John Entsminger succeeded Mulroy as the head of SNWA.

Entsminger is broad-shouldered and avuncular, a lawyer with a crewcut who radiates a masculine whimsy, decorating his corner office at SNWA's headquarters on Valley View Boulevard with a fire hydrant in Golden Knight's colors and a sculpture of a dripping kitchen tap. Almost as soon as he was placed in charge of the agency, Entsminger had realized that the third straw's usage would be limited by the pumping infrastructure in place, which wouldn't function if the level of Lake Mead dropped below one thousand feet of elevation. "I went to the engineers and said, 'We need to design a pump station that can fully utilize the capacity of the third intake.' It was blind luck, but it was at our December

2014 board meeting where the board was set to vote on funding the new pump station—during that meeting, we got a call from the engineers that the tunnel for the third intake had been completed. It was a perfect the-eagle-has-landed moment. 'You know that $800 million intake project? We just finished that, on time and under budget. We drilled a twenty-foot-diameter tunnel underneath a lake. And oh, by the way, we need another $650 million to be able to fully utilize it.'"

What's remarkable is that so much money was approved for a system whose use was almost entirely speculative. "That $650 million station only gets turned on if the surface of Lake Mead goes below 1,060 feet of elevation. So, we're spending a billion and a half dollars on a capital project that we don't know if we're ever going to need," Entsminger said. But by the time the pump station became operational, in April 2020, the turn-of-the-century drought that had forced Las Vegas and the rest of the Southwest to completely revolutionize their approach to water management had turned into a two-decade-long aridification episode, the most severe megadrought to hit the region in at least twelve hundred years. In 2022, Mead fell below 1,060 feet of elevation and the new pumps switched on. "As we're sitting here today," Entsminger said, "that intake is supplying 90 percent of the water to 76 percent of Nevada's population. We haven't always been *that* far ahead, but we got that one done."

Between 1980 and 2022, Clark County jumped from being the ninety-eighth-largest local jurisdiction in the country to ranking eleventh. For Phoenix to make a similar jump in the mid-twentieth century, it required the construction of numerous dams, not to mention the Central Arizona Project, the 336-mile aqueduct that conveys water to the city from the Colorado River. In contrast, Las Vegas's transformation into a metropolis happened without the addition of any new water resources. Not that Pat Mulroy didn't try—for years she pursued a controversial bid to pump groundwater in rural Nevada and funnel it to Las Vegas, a scheme she said would provide a "safety net if the river really goes south," but which was ultimately derailed by lawsuits from a broad array of ranchers, environmentalists, and Indian tribes.

Like it or not, since the turn of the twenty-first century Las Vegas has largely lived within its means, tightening its belt further and further as the environment has become drier and drier.

Close to half of the basins that supply freshwater across the United States may be unable to meet demand in fifty years. Already, so much water has been pumped out of the aquifers of western Kansas that the yields from the region's cornfields have plummeted; in the rapidly expanding suburbs of Washington, D.C., groundwater is declining so quickly that some communities could run dry by 2030, forcing local officials to explore building a desalination plant in Chesapeake Bay. As the rainfall that the eastern half of the country is reliant upon grows more and more irregular, these problems will only spread further, imperiling the agricultural bread basket of the Great Plains, along with cities from Omaha to Tampa Bay.

With water scarcity sure to become a pervasive feature of life all over the United States in the twenty-first century, Las Vegas represents a compelling vision of how to decouple economic and demographic growth from water consumption. But is the rest of the country ready to make the massive early investments and sacrifices that were required to achieve that decoupling? Or will conservation eventually be sidelined as demand grows for more of the massive, environmentally disruptive infrastructure projects that made the urban explosion of the Southwest possible in the first place?

Apache Junction sits at the foot of the Superstition Mountains on the far eastern edge of the Valley of the Sun. The community wasn't incorporated until 1978; even then, its population remained almost exclusively composed of leisure ranchers and the snowbirds who drove RVs to its palm-tree lined campsites in the winter, hoping to nab a spot with a decent view of the Superstitions' soaring crenelations of breccia and welded tuff. But as metropolitan Phoenix expanded, the frontier of the closest suburb, Mesa, gradually merged with that of Apache Junction. Soon, speculators were daydreaming about the flatlands south of the Superstitions.

In 2006, ASU's Morrison Institute issued a report laying out some options for developing the 275-square-mile region, which it dubbed Superstition Vistas and stated was roughly the same size as L.A.'s San Fernando Valley, thus providing enough room to accommodate nine hundred thousand people. The Great Recession put the kibosh on the Morrison Institute's grand plans, and the area south of the Superstitions remained undeveloped for a decade, until the state began auctioning off land again and Apache Junction annexed the eight square miles nearest to the town's southern edge.

Chip Wilson moved to A.J. in the early 1990s after two decades in the Air Force, drawn to the area because it was an ideal spot to keep horses. When I visited the city hall, he was wearing a cowboy hat and an official name tag that read "Chip Wilson: Mayor." After taking a drink from a large, insulated cup from Circle K, he declared, "Growth is going to be managed." This mainly entailed protecting what Wilson called "our view-shed" of the surrounding mountains, which would supposedly be endangered by building anything more than a few stories tall in the existing city.

South of town, however, developers faced far fewer limitations. "It's going to be much more modest than nine hundred thousand, that's probably three hundred years in the future," A.J.'s city manager, Bryant Powell, told me. "But we're certainly dipping our first toe in the water." In 2022, plans were announced for a pair of developers to build ten thousand homes on some of the recently annexed land over the next decade, enough to roughly double A.J.'s population from thirty-eight thousand to more than seventy thousand. "From zero to ten years out, we have a really great, clear line. Ten to twenty, we don't know what the market will be, we don't know if there will be a recession, but if there's more development it would be in the boundaries that are now within the city. Beyond that, I don't know."

The Valley of the Sun is one of the few places in America where merely doubling a city's population in ten years counts as growing at a deliberate pace. Nevertheless, part of the reason for A.J.'s relative restraint is uncertainty about how much water the city will have access to in the future. The city's groundwater is supplemented by deliveries from the Central Arizona Project. But between a state law that requires developers to demonstrate that every house they build that relies on wells has a secure source of groundwater for the next hundred years and recent cuts to Arizona's allocations from the Colorado River, building out Superstition Vistas will require A.J. to bolster its current reserves. That makes the city a prospective buyer on the emerging market for water in Arizona, which mostly serves to transfer rights from agricultural users to municipal ones.

Much of the newfangled water market is being facilitated as a conservation measure, given that 80 percent of the Colorado River's water is used by farms. Just the basin's alfalfa crop consumes significantly more water than the tens of millions of people who live in its cities. So-called demand management programs abound, where farmers are paid to cede their water rights to an urban utility. In 2003, for example, the onset of the drought forced California to figure out how to stop overdrawing its allocation of the Colorado. All eyes turned to the Imperial Irrigation District, which is the largest single river user. IID is entitled to more water than the entire state of Arizona, which it diverts to a half-million acres of farms and feedlots between the Salton Sea and the Mexican border. In the early 2000s, Imperial agreed to fork up to two hundred thousand acre-feet of water over to San Diego's utility for the next seventy-five years, while the Palo Verde Irrigation District, just to the north, chipped in another hundred thousand acre-feet to municipalities across southern California.

Demand management requires farmers to do one of three things: irrigate more efficiently, change the crops they plant, or fallow farmland entirely. In one study of corn fields in the Imperial Valley, switching from flooding the furrows between rows of corn to a drip system saved 2.2 acre-feet of water per acre. Given the Imperial Valley's massive size, widespread adoption of drip irrigation could free up a tremendous amount of water. Meanwhile, a 2015 analysis from ProPublica found that if Arizona's cotton farmers switched their fields entirely over to wheat, they would save almost double the amount of water used by the entire Tucson metropolitan area without changing anything about how they irrigate. But that strategy also requires financial inducement, given how lucrative thirsty crops like cotton and alfalfa can be. Hence the neat solution of simply paying farmers not to farm. While straightforward, fallowing requires willing partners and could disrupt agricultural markets if it becomes too widespread.

The complexities of agricultural demand management have pushed Indian tribes to the forefront of the most recent rounds of negotiation over the Colorado River, as Indigenous nations often have senior rights to more of the Colorado than they can use (that is, once those rights are formalized through litigation). In 2023, the Gila River Indian Community of Akimel O'otham and Pee Posh, which occupies a broad stripe of land south of metropolitan Phoenix, received $150 million from the federal government to leave their annual allotment of 125,000 acre-feet in Lake Mead for the following three years, helping to slow the drain of the reservoir.

I talked to Amelia Flores, the chair of the Colorado River Indian Tribe, shortly after she testified in support of a bill on Capitol Hill that would allow her tribe to broker similar agreements with municipalities in Arizona over its 650,000 acre-feet entitlement. The tribe is a confederation of Hopi, Navajo, Mohave, and Chemehuevi peoples whose land straddles both sides of the river in California and Arizona, and Flores described how Anglo settlers had first come to the region after gold was discovered in the La Paz Valley. Some five thousand people descended over the course of 1862, almost all of them vanishing once the boom went bust. "Leasing will strengthen our sovereignty over our water," Flores said. "Water is the new gold. We want to participate—everybody has to be part of the solution; it has to be everybody to save the life of the river."

Whether brokered with an irrigation district or a tribe, it's these kinds of transfer agreements that allowed urban utilities to stay ahead of the falling levels of Lake Mead over the first two decades of the twenty-first century. But with overall supply so critically low, need is growing for a more lasting solution: using less water. The most ecologically friendly way to do that while maintaining economic growth is through water recycling projects like those in Las Vegas. Those projects, much like demand management schemes, don't come cheaply. The Metropolitan Water District of Southern California, which provides for the 19 million people in greater Los Angeles and San Diego, is spending more than $3 billion on an advanced recycling plant. Phoenix is pursuing a similar project, as are smaller municipalities like South Jordan, Utah, and Rio Rancho.

"We can't develop all the land to the south because we're going to run out of our guarantee of water," Apache Junction's Chip Wilson told me. Bryant Powell agreed. "We believe we have great planning done for our current residents and for those four-square miles. Past that? We've got to figure it out. Does the water come from the Salt or the Verde? Is it from direct or indirect potable reuse that's already being done in California?"

Whatever the case, for cities like Apache Junction to thrive in the future, it's clear they'll have to adopt a far more considered approach to water management than what was practiced across the region in the first decades after the Colorado River Compact was adopted. While the completion of the Hoover and Glen Canyon Dams initially created the conditions for the Southwest to, as Herbert Hoover himself had put it, "come into its magnificent heritage of power and life-giving water," the twenty-first century has demonstrated just how grievously arid America has overshot its natural limitations.

In the winter of 2022, after the level of Lake Mead dropped to a lower level than had been measured at any point since the Hoover Dam was completed in 1936, I traveled to Las Vegas to sit in on the annual conference of the Colorado River Water Users Association. After all, where better than the halls of Caesars Palace to try to understand how the leaders charged with securing the Southwest's future were adjusting to the new reality of a river exhausted by our rapacious thirst? It only sweetened the pot that exactly one hundred years has passed since the original compact was signed.

On the first day of the conference, Brenda Burman, the incoming general manager of the Central Arizona Project, offered a history lesson. She stood at a rostrum in one of the casino's upstairs ballrooms, golden wallpaper and a pair of terra-cotta-colored columns at her back. "We've been in drought for twenty-three years," she said, pulling up a graph tracking the combined storage of Lake Powell and Lake Mead on her PowerPoint. "Those first four years of the 2000s were some of the worst hydrology we've ever seen. We lost 50 percent of our storage."

Burman described how the sudden transition to lean times had forced the seven states of the Colorado Basin to work with the Department of the Interior on a set of interim guidelines for managing the reservoirs in 2007, which was followed by an even more aggressive Drought Contingency Plan in 2019. That agreement kicked in when Lake Mead dropped below a third full in 2021, forcing severe water cuts that fell disproportionately on Arizona (at the time of the conference, the Grand Canyon State was receiving only 82 percent of its usual allocation—the next year, that number dropped to 79 percent). Meanwhile, California was receiving its full share of water, the consequence of a compromise brokered in the 1960s when the Golden State agreed to drop its opposition to the Central Arizona Project so long as its more arid neighbor assented to its demand that California's 4.4-million-acre-foot allocation go untouched in times of shortage.

Over the previous summer, the commissioner of the Bureau of Reclamation, Camille Touton, had asked all the states in the basin to agree to a plan to stabilize the reservoirs in just two months, but negotiations stalled almost as soon as they began. "We didn't make it," Burman said. "It's really hard to come up with 2 to 4 million acre-feet in sixty days." The federal government's response was to set in motion an official readjustment to the current guidelines, a process that would take just under a year to accomplish and that it hoped would include a consensus proposal for drawing use of the river down to a sustainable level. The first round of comments on that new plan were due the following week, which explained why, as Burman put it, "There's a lot of work being done in the hallways right now."

Over the three days of the conference, the various panels on "Adapting to the New Normal" and "Trades-Offs and Turbulence" proved to be as much a sideshow to the real work of guaranteeing water for everyone in the basin as the exhibition hall, where serious-looking men in polos sold pipes to agricultural engineers and enthusiastic young environmentalists handed out magnets shaped like endangered fish. While the bulk of the guests mingled on the plush carpet or squinted through the tinted windows at the statuary down in the Garden of the Gods Pool Oasis (closed for the winter), a handful of state and federal officials were holding a succession of consequential private meetings. The only public insight into those deliberations came on the second day, when the seven lawyers and engineers appointed by their governors as official delegates in all interstate Colorado River negotiations were impaneled for a public Q&A.

The moderator was Jeffrey Kightlinger, the retired general manager of the massive utility that provides water to metropolitan Los Angeles and San Diego. "These seven folks have the incredibly difficult, semi-impossible task of setting the stage for the next hundred years," Kightlinger said, before opening up the floor to the representatives from California, Arizona, and Nevada, suggesting that, since those states use three-quarters of the Colorado's water, getting overall consumption in line with a river that's only delivering 12 million acre-feet every year (as opposed to the 16.5 million assumed by the compact) meant that "the math kind of says the bulk of these cuts—maybe three-fourths, maybe more—has to come from the Lower Basin."

True to form, John Entsminger offered the bluntest assessment. "Yes, the Lower Basin is going to have to take the lion's share of the reductions," he said, even as he cautioned the crowd that the 2 to 4 million acre-feet of cuts the feds were calling for would hardly allow the basin to reach sustainability. "Our modeling shows that the cuts to the Lower Basin might be 6 million acre-feet in 2025"—an amount that, under the original compact, would leave just 1.5 million acre-feet for the three states, less than half of what the Imperial Irrigation District alone is entitled to every year. There was simply no realistic way to get to such a

low number without a massive shift in priority between agricultural and urban users.

Though California's Peter Nelson acknowledged, "We can't
continue to mine storage out of Lake Mead," he defended his
state as having already "stepped up" with its own program to
conserve some four hundred thousand acre-feet of water. Only
a massive infusion of new funds for demand management programs would make real reform possible, given that the rest of
the basin was effectively asking Southern California's growers to
completely forgo the $3.9 billion in revenue they generate every
year. Nelson's point was clear: there was nowhere near enough
money on the table to convince California to abandon its hard-
won priority right to the river.

If the basin failed to reach a consensus on how to diminish
consumption, it risked litigation—a nightmare scenario given
that interstate water lawsuits usually take decades to resolve.
There was, however, a third option, one that Arizona Senator
Mark Kelly floated during his address on the final morning of the
conference. "Augmentation projects that were once dismissed as
too ambitious, like large-scale desalinization plants and import-
ing water from other basins. Now these are ambitious ideas—I
get that. But they're no more ambitious than the Hoover Dam or
the Glen Canyon Dam when they were conceived."

Kelly's repeated use of the word *ambitious*, I later learned, was
a way of shedding an aspirational light on ideas that would carry
a mammoth cost to both taxpayers and the environment. One
water expert pointed me to a report stating even a small desalination plant in the Sea of Cortez would cost more than $70 billion
to build, require an exorbitant quantity of fossil fuels to operate, and destabilize the delicate ecosystem of the gulf's shallow
waters. Even then, that size of plant could only provide about one
hundred thousand acre-feet of water every year. Expanding such
a facility to the scale where it could actually solve the Southwest's
water shortfall would require multiplying the costs—financial,
environmental, ecological—over and over again.

After his speech, I asked Kelly if promoting desalinization
or the construction of a lengthy aqueduct to a more plenti-

ful watershed (the Missouri? the Mississippi?) risked detracting focus from the conservation measures already being pursued around the basin. "It doesn't help us solve the immediate crisis," he acknowledged. "But I think it's smart to think long term. We have 40 million people that rely on this river—I'd like to see that number go up. It's expensive and the politics behind it are often challenging, but moving water or large-scale desalinization are engineering problems we can solve."

Two days after Kelly's speech at Caesars Palace, the Howard Hughes Corporation announced that it had broken ground on a thirty-seven-thousand-acre development called Teravalis in the far West Valley of metropolitan Phoenix, which is separated from the rest of the region by the White Tank Mountains. When fully built out, the project would include over three hundred thousand residents and 55 million square feet of commercial space, making it Arizona's "largest master-planned community."

Right on cue, Arizona's Division of Water Resources released a report a few weeks later showing that the far West Valley lacked the necessary groundwater reserves to support the massive amount of development that was planned there. In modeling the next hundred years of consumption, the state found "a total unmet demand" of 4.4 million acre-feet of groundwater. Suddenly, projects like Teravalis were thrown into jeopardy: not only would all the proposed developments in the far West Valley overdraw aquifers by as much as eighty-four thousand acre-feet per year—twice as much groundwater as the entire city of Albuquerque uses annually—but the homes that *had already been built* could run out of water well before 2100.

Teravalis was slated for the north end of Buckeye, a former agricultural settlement that only transitioned from being a town to a city in 2014. The development was only one component of a plan that would transform Buckeye from an exurb of around one hundred thousand people—where subdivisions still rubbed shoulders with feedlots and alfalfa farms—to a standalone city of 1.5 million. Already, Buckeye had annexed more than four hundred square miles of land, giving it a bigger paper footprint than San Diego.

By the summer of 2023, it became clear that it was not just far-flung communities like Buckeye that would be forced to finally contend with the limits of the Sonoran Desert. The newly elected governor, Katie Hobbs, released a second report that covered the entire Phoenix region, demonstrating a potential 4 percent shortfall in groundwater over the next century. Those findings put the metropolis of 5 million people out of compliance with a 1980 law requiring all proposed residential developments to prove an assured one-hundred-year supply of water before they can be built.

In a news conference announcing a "pause" on all new residential construction that relied on groundwater—an echo of the moratorium Pat Mulroy had put on new buildings in Las Vegas back in 1991—Hobbs was careful to highlight the limited scope of her order, making clear that it would not hinder the construction of any new industrial projects, like the $40 billion microchip factory that President Biden had visited Phoenix to tout the year before. "This pause will not affect growth within any of our major cities where robust water portfolios have been proven to cover current and future demands," Hobbs said. "I cannot emphasize that enough."

Driving through Buckeye, it seemed obvious to me that the recent flurry of research and regulation was too little, too late. The future site of Teravalis was separated from I-10 by a ten-mile plain of xeric sand. Off to the west, a few roofs of homes built in the widely spaced, dusty lots of the so-called West Phoenix Estates were visible from the road. For the next two miles there was nothing, until I spotted a line of palm trees indicating the entrance to Tartesso, a marginally older master-planned community that represented the first beachhead of development into this remote corner of the Valley of the Sun.

Tartesso is a D. R. Horton venture, an eleven-thousand-acre development that first kicked off in 2005. Only three thousand of the anticipated forty-one thousand homes were completed before the financial crisis, stalling construction for the next decade. Now, just like seventy miles east in Apache Junction, the suburbanization machine was back up and running. In 2020,

Tartesso had been the top-selling master-planned community in Arizona, though Cadence, in Henderson, edged it out in the national rankings.

As soon as I turned into Tartesso, I realized Buckeye was a city still operating under a twentieth-century theory of development, that as long as growth continued a society would follow. A desalination plant, a thousand-mile aqueduct—surely some sort of public works project would come along to ensure that none of the speculators building houses out here on the open plain lost their shirt. Driving through the wriggling streets lined with three-car garages and Spanish-tile roofs, I couldn't help but think about all the vacant lots in downtown Phoenix, how few apartment buildings it would take to equal the number of homes here, marooned so far from the economic heart of the Southwest in anticipation that this twenty-first-century colony was on its way to becoming an empire in and of itself.

I found the edge of arid America a little past 304th Avenue, at the corner of Indian School and Johnson. To my left, there was a gently curving road lined with palm trees and grass. On the other side, wooden hurdles marked with fading red stripes cordoned off the wide sweep of the Sonoran Desert. I pulled over in front of the barrier and walked behind it, up to where some barbed wire was strung. Looking back at Tartesso, I could see a McMansion with solar panels on the roof, a few pigeons roosting there. It seemed like a mass-produced version of the castle Franz Huning had built in Albuquerque. The people who lived out here couldn't afford a shelter nearly as opulent as Judith Chafee's Rieveschl House, yet they still wanted more space than they needed.

A dove sat atop one of the new streetlights lining Johnson Road, softly cooing to itself. The sun radiated down on two saguaros in the distance, bright enough for me to make out the drooping of one spiny arm toward the sand. Then I heard a soft buzzing. As it got louder and clarified into an engine's roar, a plume of dust engulfed all the Russian thistle slowly curing into tumbleweeds. The 4x4 zipped into view, its driver pulling back onto the pavement a dozen yards away. He was an Anglo dude with a bushy beard, and wore a black undershirt and mirrored sunglasses that fully enclosed his eye sockets. The engine slowed to a purr as he approached me on the 4x4, taking a minute to look me up and down. In the reflection of his eyes, I saw myself: shading my face with one hand, scrutinizing his community in search of an answer to what would become of my home—if the Southwest could possibly endure. Without a word, he turned his head to the sun, cranked the accelerator, and peeled off, headed back to his own private paradise.

AFTERWORD

A complex of natural cisterns known as the Hueco Tanks sits thirty-five miles east of El Paso, at the end of a state highway that took me past firework stands, supermercados, pawn shops, strip clubs, mobile home retailers, a paintball range, and a salvage yard. I finally turned left ahead of a white stucco building shaped like a UFO that was advertising "Land Priced for You" with a sun-faded Marvin the Martian doll hanging in one window. Heading up Ranch Road 2775, there wasn't much exurban ephemera, just low sagebrush and creosote. The only real tree was a lonely cottonwood, growing on the shoulder above a concrete bench. Someone had recently painted the bench fuchsia; the cottonwood's branches were bare. I rolled to a halt and turned off my engine, gazing at the bench and attending to the quiet.

A little way north, the Hueco Tanks looked like a pile of prehistoric rubble, a jumble of eroded boulders, tallus, and sediment with no obvious connection to the chocolate-colored hills beyond. Before I could explore the formation, a park ranger escorted me to a room in the adobe visitor center to watch an orientation video about the history of the landmark—a measure taken, she

explained, to help dissuade people from defacing the ancient pictographs that can be found throughout the Hueco Tanks. Inside,
an old TV was set on a rolling stand with a cow skull gathering
dust on the lowermost shelf; against the wall, there was a series
of glass cases containing more skulls—coyote, mountain lion,
songbird—as well as a taxidermy rattlesnake and roadrunner.

In the video, which looked like it had been recorded some-
time around the 2008 financial crisis, the narrator translated the
word *hueco* to "hollow," a reference to the divots and basins in
the rock formation that capture rainwater. This natural storage system allowed life to thrive here, far from the Rio Grande.
"This place has a cultural history that goes back ten thousand,
maybe twelve thousand years," a park ranger on the grainy screen
explained. "And there's reasons for that. When the rest of the
desert is dry, Hueco Tanks has water."

The three thousand or so pictographs at the Hueco Tanks
showcase the evolution of human society in this seemingly
remote corner of the Chihuahuan Desert. The earliest paintings, starting around 6000 B.C.E., are mostly abstract designs
in red: stacks of vertical lines with intervening zigzags. Three
thousand years later, human and animal forms appear, with their
painters incorporating yellow and other tones to create figurative compositions. Then, around 450 C.E., a people known as
the Jornada took up residence at the tanks. They were closely
related to the Mogollon of the steppe separating the cliff dwellings of the Colorado Plateau from the farmers of the Salt River
Valley, and their paintings speak to their cosmopolitanism. Much
of the imagery by the Jornada Mogollon featured jaguars and the
feathered serpent known to the Aztecs as Quetzalcoatl. Though
some of the intricate, geometric designs seem sourced directly
from Teotihuacan, the combination of black, red, and white paint
is more reminiscent of Zuni pottery; likewise, the emphasis on
masks seems to be a nod toward the Katsina religion that was
concurrently evolving in the riverine villages to the north, the
places the Spanish would one day call pueblos.

After I'd finished the video, I walked into the formation.
The trail was walled in by mesquite trees, their branches hanging heavy in the sun with green pods that have provided reliable

sustenance for desert peoples over the millennia. Gradually, I approached a rock wall featuring several small impressions, each of them about as large as a cupped hand. There, a name and date had been chiseled into the stone inside a neat, square outline:

H.W.E. Well
FEB. 27
1884

I clambered up onto a shelf of what the park ranger at the visitor center had described as "honeycomb rock," pausing to catch my breath in the coolness of the cave. None of the huecos here had water, so the flies gravitated to the sweat on my face instead, buzzing insistently in my ear. On the rock wall overhead, there were more names: M.F. Wayland Jul-25 1884; J.S. Ball June 18, 1858; Mary Phillips June 16th 1887. Shapes, too. A schematic of a cube; a man in a bowler hat, his nose flat and ears sharp. Only by looking very closely could I see the older artwork that had been obscured by this settler's tag. In white, a ghostly apparition of a deer; in red, the outline of a horse. Just under J. S. Ball's signature, I thought I saw a rider, the horse white but his body black, hoisting a red implement that could've been a sickle or a bow.

I guessed that the figure had probably been left by a Mescalero Apache painter, since that tribe had often camped here in the centuries after the Jornada Mogollon left around 1400. The Apache's term at the Hueco Tanks lasted until the Mexican-American War, after which point the rock-pile oasis became a stop on the Butterfield Overland Mail route, as well as a destination for westering migrants on their way to California, one of the few reliable places to find water on the 150-mile trek between the Pecos River and El Paso. The visitor center where I checked in was actually the first permanent structure at the site, built by a man named Silverio Escontrias. He ranched here until the 1950s, building more than a dozen dams in the surrounding hills to try to capture water for his cattle, too numerous to be sated by what the tanks could provide.

Since the Hueco Tanks became a state park in the 1960s, rangers have done their best to preserve the pictographs from further damage, closing the most sensitive areas and requiring guided tours to many portions of the formation. Curiously, the relics under protection at the Hueco Tanks include all those nineteenth-century names that had been carved over the ancestral artworks. Back on the trail, an informational placard made a distinction between "Historic Graffiti" and "Destructive Graffiti." While the former had to be protected to commemorate the "surveyors, Buffalo Soldiers of the 9th Cavalry, cattle drivers and explorers stopping at Hueco Tanks for water and shelter," anything "modern" was strictly prohibited.

I pressed deeper into the formation and found a shady shelf between two towering rock walls. Looking east, toward the Hueco Mountains, the emerald mesquite trees provided a startling contrast to the tawny hills beyond, the blue sky above. I drank some water, then let a fly alight on the lip of my bottle before I screwed the cap back on. We were both thirsty, and the tanks, that day, were dry. Water equals life, no matter the climate—the only complication in the Southwest is just how narrow the margins are. The rock cisterns of Acoma, the canals of Phoenix, the aguadores of El Paso, Las Vegas's artesian springs, the muddy Rio Grande.

As the climate crisis has drained away the Southwest's stockpile of Colorado River water, the so-called bathtub ring around Lake Mead has become a Paleozoic metaphor for scarcity. The last time I visited the reservoir, I too was struck by the walls of Boulder Canyon. A sharp line divided the buff and scarlet of the higher elevations from the white rock that was previously submerged, its crispness an unintentional extension of the feat of engineering that had created Lake Mead in the first place.

Given the tremendous amount of development it unleashed, the Hoover Dam represents the culmination of the nineteenth-century ambitions to colonize arid America. Since American railroads first pierced the Southwest 150 years ago, the region's residents have fully transitioned from relying on ancient stopovers like the Hueco Tanks to aqueducts hundreds of miles long, yet the same challenge persists: the desert gives little, no matter how much we ask. For a century and a half, Americans believed industrial know-how could change the fundamental ecology of the desert. That effort failed, but what remains to be seen is what we do now that recycling and conservation technologies are making it possible to return to a system of living that respects the limitations of the landscape, albeit at far greater scale.

How will we use our power? Will we seek new ways to exploit and master the desert, or commit to adaptation? For all the intermingling of cultures that has created the contemporary Southwest, a stark division remains between the restless mindset of the colonizer and that of the patient neighbor who is comfortable making do with less. We can focus on sustaining ourselves, housing each other, and making room for new migrants willing to live by the same ethos of community and environmental care. Or we can continue to emphasize economic growth at all cost, even as we know that, inevitably, the reservoirs will fill with silt and the pumps will dredge sand once there's no groundwater left. Down that path lies chaos: waves of climate refugees fleeing to the once-temperate climes of America at the same time northeastern and midwestern cities are newly contending with the same problems that far too many southwesterners still refuse to acknowledge.

Life in the desert is not new, but that does not mean that we can take the current scale of it for granted. As the twenty-first century wears on, it will become impossible to ignore the eternal truth that the earth sets the conditions, and the only sensible path for humanity is to meet those terms with humility and grace. Such an arrangement might be antithetical to the conquerors who laid claim to the continent over the course of five centuries. But on this planet that predates us and will outlive us, people are just a phase. Our duration will depend on our willingness to attend to the inherent logic of our home.

ACKNOWLEDGMENTS

Writing this book has been the culmination of an ambition I've harbored since I can remember having ambitions. So, even though I was never a particularly dedicated student, please allow me to offer my most sincere gratitude to the many teachers who have nudged me forward along the way. In Albuquerque: William Kuh, Eder J. Williams McKnight, George Ovitt, Casey Kile Citrin, Jim Linnell, and Richard Hogle. In Boston: Jay Cantor, Jonathan Wilson, Ronna Johnson, John Lurz, Ichiro Takayoshi, Margo Caddell, and Ted Simpson. In New York: Victor LaValle, Christopher Sorrentino, Heidi Julavits, Donald Antrim, Ben Marcus, Mike Harvkey, Paul Beatty, and Rivka Galchen.

I am eternally grateful to Christopher Beha for opening a door to a career in writing for me at *Harper's*, and to Jonathan Sturgeon for publishing my first few essays at *The Baffler*. It's only possible to make a living in magazines through the largesse of others, so thank you to Luke Zaleski, Christopher Cox, James Marcus, Soraya King, Camille Bromley, Kevin Lozano, Ryu Spaeth, Zachariah Webb, James Yeh, Catherine Elton, Tom Stackpole, Will Stephenson, and the many other editors who have proven to be such willing and insightful collaborators.

Levelheaded yet forceful, Jonah Straus is as resolute a partner as any author could ask for in an agent. To Alec Yoshio MacDonald: I'll always appreciate you seeing some potential in my writing in the first place.

It would have been impossible for me to turn my murky cloud of memories, research, and interviews into something lucid without the patient guidance of David Treuer, whose confidence in the worthiness of this project has never wavered. Pantheon is lucky to have him, and the same goes for Shanna Milkey, Rose Cronin-Jackman, Linda Huang, Sam Chivers, Amy Stackhouse, and Kevin Bourke. A particular shout-out to Amara Balan and Lisa Lucas, who were vital early partners in getting *American Oasis* into fighting shape.

I am grateful to the dozens of southwesterners who kindly made time to speak with me for this book, especially those who also provided introductions or recommended books to read and sites to visit. Bob Moore, Scott Dickensheets, Kellen Braddock, Raquel Gutiérrez, Christopher Domin, and Michelle Malonzo were particularly helpful in getting me oriented in their respective cities. Special thanks to Kathy McGuire and Demion Clinco for organizing my tour of several of Judith Chafee's most remarkable buildings. As I was finishing the first draft, Rachel Monroe graciously invited me to cat-sit for her in Marfa, which proved to be an absolutely necessary retreat. Amy Reichenbach made me look like an author. Mathis Clément did an impeccable job fact-checking this book, and any mistakes that remain are my own. The map reprinted in the opening pages was a gift from Laura Swanson, supreme New Mexican and eternal homie. Thanks, pal.

Of course, I would never have been able to find my way as a writer without the unequivocal support of my parents—Mom and Dad, it's rare to be someone who seeks to pursue a life in the arts without a cacophony of second-guessing and well-meaning suggestions to try a more straightforward path, and I do not for a moment take your belief in me for granted. Likewise Nathan, Liz, and Simone, the consummate cheerleader. Mark and Ann, the stability I needed to complete this book stems from your generosity—thank you.

To Tess: You remain my first and best reader, my champion, my source of constancy and ever-renewing joy. Next one's for you.

Lastly, a shout to all my Burqueños, whether I talk to you

every week, haven't seen you in years, or somewhere in between. The Academy kids, Boris, Allie, Vlado, Carrie, Gabri, John, Devon, Anna, Claire, Julie, Terri, Julia, Carly; the Manoa crew, Lauren, Dani, Matt, Skye, Annie, Kate, Juli, Elsa, Katy, Summer; and, of course, my Sandia Park soul brother, Ryder. Órale!

NOTES

vii "In case of building": Felipe II, *Transcripción de las ordenanzas*, 38 (translation is my own).

vii "I am crying from thirst": "I Am Crying from Thirst," in Dodge and McCullough, *Wah'Kon-Tah*, 68.

PREFACE

xv first Spaniard to chronicle: Cabeza de Vaca, *Adventures in the Unknown Interior of America*.

xvi "clean, blank page": Smythe, *The Conquest of Arid America*.

xvii magazines dismiss Albuquerque: Julia Felsenthal, "Three Days at Los Poblanos, the Dreamy Albuquerque Farm That Might Make You Rethink Your Life," *Vogue*, August 12, 2015.

xvii "if not compelled": Hodge, *Texas Blood*, 222.

xviii trio of researchers: E. Hertz, T. Haywoode, and L. T. Reynolds, "The Sunbelt Syndrome: Radical Individualism at the End of the American Dream," *Humanity & Society* 14, no. 3 (1990): 257–79.

xix family of Lebanese immigrants: Cartwright, *Dirty Dealing*, 15–27.

xix the Black Striker: Ibid., 116–18.

xix high-stakes blackjack players: Ibid., 68–70.

xx collapse of the Chagra family: Ibid., 112–20.

MARROW

3 "Its rank is a mountain": Layli Long Soldier, "Ȟe Sápa, One," in *Whereas*, 6.

7 four survivors: Cabeza de Vaca, *Adventures in the Unknown Interior of America*.

7   vertical cities: "Chaco Culture," UNESCO World Heritage Conven-
    tion, www.unesco.org/; "Ancestral Pueblo People and Their World,"
    Mesa Verde National Park, 2019, npshistory.com/; "The Center of
    Chacoan Culture," Chaco Culture National Historical Park, 2017, www
    .npshistory.com/; "The Museum Collections of Chaco Culture National
    Historical Park," National Park Service Museum, www.nps.gov/.
8   fifty-year drought: deBuys, *A Great Aridness*, 84–87.
8   ornate Katsina dolls: Roberts, *Pueblo Revolt*, 36–40; James Blake Wiener,
    "Kachina Cult," World History Encyclopedia, January 16, 2016, www
    .worldhistory.org/.
8   ample view of the scrublands: Roberts, *Pueblo Revolt*, 55–59.
9   base of operations: Hoig, *Came Men on Horses*, 69; Gibson, *El Norte*, 63.
10 storming of Hawikuh: Kessel, *Kiva, Cross, and Crown*.
10 province known as Tiguex: Bolton, *Coronado*, 184–86; Hoig, *Came Men
    on Horses*, 72; and Roberts, *Pueblo Revolt*, 28.
11 largest of these settlements: Erik Reed, "Southwestern Indians in Coro-
    nado's Time," National Park Service, *Region III Quarterly* (July 1940).
11 eighty thousand people: Roberts, *Pueblo Revolt*, 29.
11 Coronado's second-in-command: Flint, *Great Cruelties Have Been
    Reported*, 336–37; Hoig, *Came Men on Horses*, 73.
11 scattering across the valley: Simmons, *Albuquerque*, 33; Bolton, *Coronado*,
    92–93.
11 the days grew colder: Hoig, *Came Men on Horses*, 79–80.
12 black smoke billowed: Bolton, *Coronado*, 210–11.
12 sign of the cross: Hoig, *Came Men on Horses*, 81.
14 Juan de Oñate: Simmons, *Last Conquistador*, 14–26; Timmons, *El
    Paso*, 12.
14 Under his command: Roberts, *Pueblo Revolt*, 72–73.
15 muddy river: running high: Timmons, *El Paso*, 13.
15 capital of the colony: Roberts, *Pueblo Revolt*, 75–76; Treib, *Sanctuaries of
    Spanish New Mexico*.
16 350-foot climb: "Acoma Pueblo: Ancient City in the Sky," National Park
    Service, www.nps.gov/.
16 the Acoma were incredulous: Roberts, *Pueblo Revolt*, 85–87; Gibson, *El
    Norte*, 68.
17 another abandoned pueblo: Federal Writers' Project, *New Mexico*, 188–
    89; Fugate, *Roadside History of New Mexico*, 113.
17 a thousand settlers: Gibson, *El Norte*, 69.
18 enslaved Puebloans: Reséndez, *The Other Slavery*; Simon Romero,
    "Indian Slavery Once Thrived in New Mexico. Latinos Are Finding
    Family Ties to It," *New York Times*, January 28, 2018, www.nytimes.com/.
18 form of devil worship: Roberts, *Pueblo Revolt*, 116.
18 smallpox outbreak and drought: Ibid.
18 supposedly colluding: Ibid., 117–24.

18 collective action: Roberts, *Pueblo Revolt*, 126; Gibson, *El Norte*, 74; Sando, *Po'pay*, 17–18.

19 "ripe cultigens": Sando, *Po'pay*, 85.

19 meeting in secret: Ibid., 21–29.

20 figure of Saint Francis: Simmons, *Albuquerque*, 42.

20 naked through the mud: Roberts, *Pueblo Revolt*, 15–16.

20 body on a pyre: Eskeets and Kristoforic, *Send a Runner*, 34.

21 army of five hundred: Roberts, *Pueblo Revolt*, 22–27.

21 not happy times: Ibid., 166.

22 executed by firing squad: Ibid., 172–79.

22 "forced to adopt Christianity": Daniel Chacón, "Debate Heats Up over Annual Fiesta's One-Sided Story," *Santa Fe New Mexican*, September 22, 2015.

23 "getting laid to rest": Daniel Chacón, "Fiesta Drops Divisive Entrada Pageant in Santa Fe," *Santa Fe New Mexican*, July 24, 2018.

23 grim, two-year campaign: Roberts, *Pueblo Revolt*, 205–10.

23 six Zuni villages: "Hawikuh and the Zuni-Cibola Complex," National Park Service, www.nps.gov/.

23 more as military allies: Douglas W. Richmond, "The Climax of Conflicts with Native Americans in New Mexico: Spanish and Mexican Antecedents to U.S. Treaty Making During the U.S.-Mexico War, 1846–1848," *New Mexico Historical Review* 80, no. 1 (2005), www.digitalrepository .unm.edu/.

24 ducal title: Simmons, *Albuquerque*, 81–83.

25 client state of the Comanche: Hämäläinen, *Comanche Empire*, 32.

25 "the land is fertile": Ibid., 85.

25 "in the hands of fortune": Ibid., 77.

26 "rather ruinous aspect": Simmons, *Albuquerque*, 164.

26 "the dirtiest hole": Ibid., 168.

26 serve out the terms: Reséndez, *The Other Slavery*.

26 number of Genízaros: Simon Romero, "Indian Slavery Once Thrived in New Mexico. Latinos Are Finding Family Ties to It," *New York Times*, January 28, 2018.

26 "rod of terrorism": Locke, *Book of the Navajo*, 278.

27 dismantling of Dinétah: Ibid., 351.

27 five thousand peach trees: Mercer Cross, "Ex-Fighter Pilot Helps Navajo Dream Come True with Trees Program," *Los Angeles Times*, January 3, 1993.

27 "cease to exist or move": Locke, *Book of the Navajo*, 356.

27 One in five of the Navajo: Jennifer Davis, "Naaltsoos Sání and the Long Walk Home," Library of Congress Blogs: In Custodia Legis, June 18, 2018, www.blogs.loc.gov/.

27 decimated by cutworm: Eskeets and Kristofic, *Send a Runner*, 158.

27 "we loved it so": Ibid., 178.

28 his presidential campaign: "1844 Democratic Party Platform," May 27, 1844, American Presidency Project, University of California, Santa Barbara, www.presidency.ucsb.edu/.

28 "the immediate and connecting territory": James Polk, "Fourth Annual Message to Congress," speech, Washington, D.C., December 5, 1848, University of Virginia Miller Center.

28 fortune in the new world: Ricardo S. Gonzales, "The L. & H. Huning Mercantile Company: A Case Study of Mercantile Conquest in the Rio Abajo Region of New Mexico, 1848–1880," 2017, www.digitalrepository .unm.edu/.

29 "I accepted": Simmons, *Albuquerque*, 154.

29 families having fled south: Ibid., 129–30, 164.

29 occasional market days: Ibid., 170.

29 garrison in Albuquerque: Ibid., 155–56.

29 sawmill for the timber: Albuquerque Historical Society, "U.S. Territorial Economy, 1846–1912," 2008, www.albuqhistoc.org/.

29 began buying land: Gonzales, "The L. & H. Huning Mercantile Company"; Simmons, *Albuquerque*, 218.

29 leading citizen of the town: Simmons, *Albuquerque*, 138.

30 scrutinized their maps: Ibid., 211–15.

30 New Mexico Town Company: Gonzales, "The L. & H. Huning Mercantile Company."

30 shrugged their shoulders: Fergusson, *Albuquerque*, 66–67.

31 solid mahogany bar: Ibid., 42.

31 Italianate manor: Bannerman, *Images of America*, 35; Simmons, *Albuquerque*, 279.

31 posted a sign: Simmons, *Albuquerque*, 280.

31 "government of the white man": John Calhoun, "Speech on the War with Mexico," Washington, D.C., January 4, 1948, Papers of John C. Calhoun.

32 citizenship itself was invalidated: Gibson, *El Norte*, 222.

32 regain the franchise: Maggie Toulouse Oliver, "2022 Margaret Chase Smith American Democracy Award Submission: Miguel H. Trujillo," New Mexico Secretary of State, July 22, 2022, www.sos.nm.gov/.

32 "pure-blood or Castilian": John Nieto-Phillips, "Spanish American Ethnic Identity and New Mexico's Statehood Struggle," in Gonzales-Berry and Maciel, *Contested Homeland*, 113.

32 "ignorant and degraded": Ibid., 119.

32 "ignorant of our laws": Ibid., 120.

33 "Cortez, Pizarro, and Alvarado": Ibid., 122.

33 "No other blood": Ibid., 125.

33 "But it is *not* done": Smythe, *Conquest of Arid America*, xiii.

33 "agricultural and horticultural district": Ibid., 231.

33 "growth of the white population": Ibid., 247.

34 "semi-bustling American towns": Lummis, *Land of Poco Tiempo*, 4–6.
34 "a romance and a glory": Ibid., 9.
34 "simpler agricultural tasks": Smythe, *Conquest of Arid America*, 257.
35 formally incorporated: Simmons, *Albuquerque*, 233.
35 New Town also absorbed: Ibid., 234, 338.
35 began paying painters: McLuhan, *Dream Tracks*, 19.
35 "ancient Indian pueblos": Ibid., 22.
35 Wealthy travelers flocked: Ibid., 41.
35 In their silent movie: William and George Allen, dirs., *Primitive Indians of the Painted Desert*, 1920s, collected in the American Indian Film Gallery, University of Arizona, www.aifg.arizona.edu/.
35 Hotel Alvarado: Bannerman, *Images of America*, 51–62.
37 Sellers's script: Fergusson, *Albuquerque*, 69–70.
37 greet the dignitaries: Ibid., 71–72.
38 so-called lunger: Jake W. Spidle Jr., "An Army of Tubercular Invalids: New Mexico and the Birth of a Tuberculosis Industry," *New Mexico Historical Review* 61, no. 3 (1986), www.digitalrepository.unm.edu/.
38 "as if he were Mercury": David Katz, "Nathan Glassman Starred as an Athlete over Fifty Years Ago," author's collection.
38 Monte Vista Addition filled in: David Kammer, "Albuquerque's 20th Century Suburban Growth," New Mexico Office of the State Historian, November 3, 2013, www.newmexicohistory.org/.
38 Old Town was in decline: Benny Andres, "La Plaza Vieja," in Gonzales-Berry and Maciel, *Contested Homeland*, 239–68.
39 Kirtland Air Force Base: DeMark, *Essays in 20th Century New Mexico History*, 112; Luckingham, *Urban Southwest*, 76.
39 acequias that had fed farms: Gonzales-Berry and Maciel, *Contested Homeland*, 251.
39 "rundown and useless": Ibid., 255.
40 handicrafts and trinkets: Ibid., 252.

DESTINY

43 People talk about environment: McPhee, *Encounters with the Archdruid*, 191.
43 cottonwoods, tamarisks, and willows: Bruce Berger, "Phoenician Shipwrecks," in *A Desert Harvest*, 61–78.
43 clutching onto a rope: Paul Messinger, "Throwback Thursday: Canal Skiing Was Popular Activity," *Arizona Republic*, September 17, 2015, www.azcentral.com/.
44 its footprint exploding: Courtney Columbus, "Changing Landscapes, Rising Temperatures in Maricopa County," Arizona PBS, April 28, 2017, www.cronkitenews.azpbs.org/.
44 extreme urban heat island: Anthony Brazel, "Urban Heat Island Affects

Phoenix All-Year Round," *Arizona Republic*, September 22, 2007, www
.sustainability-innovation.asu.edu/.

44   summer highs in the city: Audrey Jensen, "Record-Breaking August
      Cements Summer 2020 as Hottest in Phoenix History," *Arizona Republic*,
      September 5, 2020, www.azcentral.com/.

45   only reference point: Thomas Cooper, "*Arizona Highways*: From Engi-
      neering Pamphlet to Prestige Magazine," master's thesis, University of
      Arizona, 1973, 52.

45   "thirty-percent more house": Allen Reed, "Dream Homes by the
      Dozen," *Arizona Highways* (September 1954).

47   planting corn in the valley: "The Hohokam," Arizona Museum of Natu-
      ral History, www.arizonamuseumofnaturalhistory.org/.

47   corruption of an O'odham word: "The Ancestral Sonoran Desert Peo-
      ple," Casa Grande Ruins, National Park Service, February 10, 2021,
      www.nps.gov/.

47   Small villages were formed: Gregonis and Reinhard, *Hohokam Indians of
      the Tucson Basin*, 8–9.

48   precipitous decline: "Pieces of the Puzzle," Archaeology Southwest,
      www.archaeologysouthwest.org/; "Hohokam Culture," National Park
      Service, August 14, 2017, www.nps.gov/.

48   instability of the climate: deBuys, *Great Aridness*, 84–85, 187.

48   "Yours Truly": Trimble, *Roadside History of Arizona*, 180–95.

48   distant Army garrison: Stephen Shadegg, "The Miracle of Water in the
      West," *Arizona Highways* (July 1942).

48   hauling it fifteen miles: Geoffrey P. Mawn, "Promoters, Speculators, and
      the Selection of the Phoenix Townsite," *Arizona and the West* 19, no. 3
      (1977): 207–24, www.jsotr.org/.

49   profit off the prospectors: Shelby Carr, "Ten Fascinating Facts About
      Jack Swilling, the Founder of Phoenix," *Old West*, November 21, 2022,
      www.oldwest.org/.

49   bullet was never removed: Amy Ouzoonian, "Jack Swilling," *True West*,
      August 6, 2013, www.truewestmagazine.com/.

49   "render me almost crazy": Linda Swilling Regan, "Jack Swilling of Ari-
      zona," American Pioneer & Cemetery Research Project, March 2009.

49   itinerant English nobleman: Niki D'Andrea, "Duppa House," *Phoenix
      New Times*, July 7, 2010, www.phoenixnewtimes.com/.

49   tentative moniker: Mawn, "Promoters, Speculators"; Clay Thompson,
      "Was Phoenix Once Called Pumpkinville?," *Arizona Republic*, Janu-
      ary 19, 2015, www.azcentral.com/.

50   amazed by their farms: William Emory, "Report on the United States
      and Mexico Boundary Survey," submitted to the Department of the Inte-
      rior, 1857, 116.

50   in just three years: Logan, *Desert Cities*, 47, 64.

50   "sleepy, semi-Mexican features": Ibid., 85.

50   "the glory of Arizona": Smythe, *Conquest of Arid America*, 249.

51   "much less numerous": Ibid., 256.

51   "by Anglos, for Anglos": Luckingham, *Minorities in Phoenix*, 18.

51   "best kind of people": Needham, *Power Lines*, 40.

51   knocking out a railroad bridge: Sheridan, *Arizona*, 208–10.

52   "saved and used for irrigation": Reisner, *Cadillac Desert*, 112.

52   most significant achievement: "A Brief History of Roosevelt Dam," Bureau of Reclamation, June 25, 2015, www.usbr.gov/.

52   reliable supply of water: Sheridan, *Arizona*, 217.

53   created a commission of delegates: "League of the Southwest Sec'y Tells of the Riverside Meeting," *Mojave County Miner*, December 30, 1921; "Author of Plan of Water Board Praises Bursum," *Albuquerque Morning Journal*, September 20, 1921.

53   their claim to the river: Kuhn, *100 Years Ago in the Negotiations of the Colorado River Compact*, 25–39.

54   "wild Indian article": Ibid., 51–54.

55   critical accounting error: deBuys, *Great Aridness*, 141–45.

55   on November 24: Kuhn, *100 Years Ago in the Negotiations of the Colorado River Compact*, 67–69.

55   "foundation has been laid": Ibid., 71.

56   pipeline of redwood: Karen Smith, "Community Growth and Water Policy," in Johnson, *Phoenix in the Twentieth Century*, 155–63.

56   "have faith": William J. Williams, "Lin B. Orme," *The Reclamation Era*, Official Publication of the Bureau of Reclamation, January 1953, 114.

56   diorama of the Roosevelt Dam: Douglas Towne, "80 Years Ago Downtown Phoenix Hosted a Wet 'n Wild Party for the Ages," Downtown Phoenix Inc., November 5, 2020, www.dtphx.org/.

57   "It will rain": Ibid.; Johnson, *Phoenix in the Twentieth Century*, 162.

57   was still a child: Cooper, "*Arizona Highways*," 71.

58   three different editors: Selnow and Riley, *Regional Interest Magazines of the United States*, 12.

58   "travel to and through": "Factual Review 1962," Arizona Highway Department, 4.

58   "poem in concrete": Raymond Carlson, "Water in the Magic Land," *Arizona Highways* (April 1941).

59   lounge in the sun: Ray Berry, "Artillery with Kentucky Windage," *Arizona Highways* blog, December 7, 2013.

59   "one finds Arizona more attractive": Raymond Carlson, "Editor's Letter: Gentle . . . the Winds of March," *Arizona Highways* (March 1946).

59   "Arizona's Indispensable Man": Don Dedera, "Walter Reed Bimson: Arizona's Indispensable Man," *Arizona Highways* (April 1973).

59   70 percent of deposits: Needham, *Power Lines*, 55–56.

59   Dwight Heard: Johnson, *Phoenix in the Twentieth Century*, 18; Sheridan, *Arizona*, 209.

59   more than 7,500 acres: Sheridan, *Arizona*, 214.

60   "Mr. and Mrs. America": Raymond Carlson, "Editor's Letter: Lazy Weather," *Arizona Highways* (June 1946).

60   all-color special edition: Robert Stieve, "A December to Remember," *Arizona Highways* blog, www.arizonahighways.com/.

60   Ansel Adams photo: *Arizona Highways* (June 1946).

60   "more real and poignant": Carlson, "Editor's Letter."

60   most renowned nature photographers: See art of *Arizona Highway* issues from March 1946, January 1953, and January 1955.

61   Valley of the Sun: Luckingham, *Urban Southwest*, 71, 102.

61   "your home base": "Phoenix: Everything Under the Sun," undated pamphlet circulated by the Phoenix Chamber of Commerce, author's collection.

61   "vacation special offer": "Enjoy a Wonderful Vacation," undated direct mail advertisement circulated by Del E. Webb Development Co., author's collection.

61   "habit forming": Inez Robb, "Grandiose 'Arizona Highways' Propaganda Imparts Ravishing, Hard to Believe Truth," *Arizona Daily Star*, February 12, 1965.

61   "handiwork of the engineer": Raymond Carlson, "Editor's Letter: Servant of the People," *Arizona Highways* (April 1961).

62   "great Southwest sun county": Tim Kelly, "The Changing Face of Phoenix," *Arizona Highways* (March 1964).

62   coined the moniker: Sheridan, *Arizona*, 268.

62   play-acting Indian: Needham, *Power Lines*, 91–92.

63   "It was all timing": Robert Nelson, "Par Tee On," *Phoenix New Times*, January 18, 2001, https://www.phoenixnewtimes.com/news/par-tee-on-6419333.

63   Sperry Rand was scouting locations: Luckingham, *Urban Southwest*, 83; Needham, *Power Lines*, 90–91.

63   output of Phoenix ballooned: Luckingham, *Urban Southwest*, 82.

64   "from his hospital bed": Joe Stacey, "Editor's Letter: Mr. Arizona Retires," *Arizona Highways* (November 1971).

64   every day of his retirement: "Raymond Carlson, Edited 'Highways,'" *Arizona Republic*, February 1, 1983; Gary Avey, "Editor's Letter: The Man Who Took Arizona to the Rest of the World . . . Raymond Carlson, 1906–1983," *Arizona Highways* (April 1983).

65   two-hundred-mile ride: Amy Long, "Hashknife Pony Express," Salt River Stories, July 31, 2013, www.saltriverstories.org/.

65   "the flavor and spirit": Cynthia Buchanan, "The Hashknife Pony Express," *Arizona Highways* (January 1988).

RECALCITRANCE

69 Only the sunlight: Abbey, *Desert Solitaire*, 135.

69 continent-spanning ties: Gregonis and Reinhard, *Hohokam Indians of the Tucson Basin*, 23–28.

70 version of the game ullamaliztli: Alexandra Witze, "The Mystery of Hohokam Ballcourts," *Archeology Magazine*, March 15, 2018, www.archaeologicalconservancy.org/.

70 four hundred thousand acres: Joseph Schmidt, "Cotton Continues to Pad Farmers' Pockets," Cronkite News, ASU Cronkite School of Journalism, www.cronkitenewsonline.com/.

70 hybridized Egyptian cotton: Julie Murphree, "'Why in God's Name Are We Growing Cotton in the Desert?,'" Arizona Farm Bureau, May 18, 2016, www.azfb.org/; Jay Mark, "Cotton's History in Arizona Began with Tempe Man," *Arizona Republic*, July 18, 2014, www.azcentral.com/.

71 "America's ugliest street": Loudon Wainwright, "Look Down, Look Down, That Loathsome Road," *Life*, July 24, 1970.

71 a famous polyglot: Wolfgang Saxon, "Kenneth L. Hale, 67, Preserver of Nearly Extinct Languages," *New York Times*, October 19, 2001, www.nytimes.com/; "Kenneth Hale, Linguist and Activist for Endangered Languages, Dies at 67," Massachusetts Institute of Technology News, October 17, 2001, www.news.mit.edu/.

71 fifteen thousand speakers: "Endangered Languages: The Full List," *Guardian*, April 15, 2011, www.theguardian.com/.

72 building a language center: "Tribal College's Tohono O'odham Language Center Is Leading Effort to Preserve and Revitalize the O'odham Language," *Runner*, August 7, 2020, www.tocc.edu/.

72 charter high school: Lauren Gilger, "Welcome to Ha:Sañ Prep, the Tucson High School Bridging O'odham Tradition with Modern Education," KJZZ, April 20, 2022, www.fronterasdesk.org/.

72 "Tucson is a linguistic alternative": Ofelia Zepeda, "Ceremony," in *Where Clouds Are Formed*, 43.

73 Spanish and Tohono O'odham farmers: Logan, *Desert Cities*, 27–29, 32.

74 "no place for a poor white man": Ibid., 38.

74 pattern of segregation: Ibid., 88.

74 barrios all over town: Otero, *La Calle*, 33–36.

75 "land of the hot dog": Ibid., 79.

75 "bad impression of the city": Ibid., 80.

75 "racial and ethnic groups": Ibid., 89.

75 "slum clearance program": Ibid., 97.

76 "a step against the tide": Ibid., 100.

76 stopped collecting trash: Ibid., 101.

76 "going to get torn down": Margaret Regan, "There Goes the Neighborhood," *Tucson Weekly*, March 6, 1997, www.tucsonweekly.com/.

76 "over the entire downtown": Ibid.

76 three thousand votes: Otero, *La Calle*, 117–18.

76 "adequate low-cost housing": Ibid., 114–15.

77 San Juan Hill: Anastasia Tsioulcas, "Revisiting San Juan Hill, the Neighborhood Destroyed to Make Way for Lincoln Center," NPR, October 7, 2022, www.npf.org/.

77 Atlanta's Summerhill: Adam Paul Susaneck, "Segregation by Design," Delft University of Technology Centre for the Just City, 2024, www.segreationbydesign.com/; Rebecca Burns, "From Streetcar Suburb to Stadium Site," *Atlanta*, June 20, 2013, www.atlantamagazine.com/.

77 Boston's West End: Anna Boyles, "The Demolition of the West End," City of Boston Archives and Records Management, March 27, 2020, www.boston.gov/.

77 Gay and Lesbian Latinos Unidos: Lydia Otero, "My Archives: 20 Years of Los Angeles' LGBTQ+ Movement," *High Country News*, March 23, 2022, www.hcn.org/.

79 "up and down arroyos": Domin and McGuire, *Powerhouse*, 21.

79 prototypical modernist architect: Margo Hernandez, "Uncompromising Architect Is Shaping Dreams," *Arizona Daily Star*, February 19, 1989.

79 educated far afield: Domin and McGuire, *Powerhouse*, 23–38.

79 "politics of New York architecture": Patterson, "At Home in the Desert."

79 unspoken prohibition: Domin and McGuire, *Powerhouse*, 38.

79 last years on the East Coast: Ibid., 40.

80 "sell honest buildings": Patterson, "At Home in the Desert."

81 four rowhouses in El Presidio: Domin and McGuire, *Powerhouse*, 68–77.

81 block was in rough shape: Nancy Sortore, "Young Architect Remodels Ancient Adobe to Serve as Office and Living Quarters," *Arizona Daily Star*, March 21, 1971.

81 fused modernist massing: Curtis, *Modern Architecture*, 637–38.

82 "best architecture in Tucson": Patterson, "At Home in the Desert."

84 "glare from the punched windows": Dan MacMasters, "A Study in the Use of Light," *Los Angeles Times*, March 30, 1975.

85 stepped library stair: Domin and McGuire, *Powerhouse*, 110–23.

85 "a lot of spatial delight": Hernandez, "Uncompromising Architect Is Shaping Dreams."

86 Farmington Road and 24th Street: Otero, *Shadows of the Freeway*, 6–11.

86 deep roots in Tucson: Ibid., 17–31.

86 "seemed like paradise": Ibid., 46–48.

87 "lived on the construction site": Ibid., 49–51.

87 "Making adobe bricks": Gutiérrez, *Brown Neon*, 52.

87 stable interior temperatures: George Austin, "Adobe as a Building Material," *New Mexico Geology* 6, no. 4 (November 1984): 69–71.

88 "combined with the adobe": Domin and McGuire, *Powerhouse*, 201–208.

88 "cool and inviting outside space": Otero, *Shadows of the Freeway*, 62.

88 150 homes in Barrio Kroeger Lane: Ibid., 53–60.

88 "drivers ran over the body": Ibid., 74.

89 seven-year-old Robert Sepulveda: "Father of Dead Boy Files Suit," *Tucson Citizen*, September 1, 1969.

89 received a settlement: Otero, *Shadows of the Freeway*, 71.

89 ran into and throughout the house: Ibid., 47.

89 "dictated the planning agenda": Lydia Otero, "Modernism Week: A Golden Age for Whom?," *Tucson Weekly*, October 22, 2015, www.tucson weekly.com/.

92 built for Oliver North: Arthur Rotstein, "Investigation May Link North, Tucson House," *Arizona Daily Star*, June 19, 1987.

92 "as little as possible": Domin and McGuire, *Powerhouse*, 227.

CHIMERA

97 In a landscape: Banham, *Scenes in America Deserta*, 44.

98 "possession, once, twice, and thrice": Fugate, *Roadside History of New Mexico*, 15; Encinias, *Two Lives for Oñate*, 45–46.

99 Indians who were allied: Timmons, *El Paso*, 18.

99 landmark was formed: Jackson Polk, dir., *El Paso's Mount Cristo Rey*, Capstone Productions and the El Paso County Historical Society, El Paso, 2007.

99 progress through the vineyards: Timmons, *El Paso*, 25–37.

100 fifteen hundred wagons: Ibid., 105.

100 industries that spread across Chihuahua: Alvarez, *Border Land, Border Water*, 71–75.

100 personally invited by the dictator Porfirio Díaz: Hernández, *Bad Mexicans*, 36.

100 El Paso Smelter: Martin Donell Kohout, "ASARCO," Texas State Historical Association, November 1, 1994, www.tshaonline.org/.

100 five preeminent madams: Trish Long, "Sin City That Was El Paso Drew Nightlife, Red Lights, Madams," *El Paso Times*, February 27, 2020, www .elpasotimes.com/.

100 early free trade zone: Samuel E. Bell and James M. Smallwood, "Zona Libre: Trade and Diplomacy on the Mexican Border 1858–1905," *Arizona and the West* 24, no. 2 (1982): 119–52, www.jstor.org/.

101 national publicity campaign: Dominique Ahedo et al., "Prohibition Stimulated Economies of El Paso, Juárez," *Borderlands* 19 (2000), www .epcc.libguides.com/.

101 "High class tourists": Engelbrecht, *Henry Trost*, 77.

101 "proms often ended in Juárez": Burciaga, *Drink Cultura*, 81–85.

102 two or three times the national average: Francisca James Hernández, "Dislocation and Globalization of the United States–Mexico Border:

The Case of Garment Workers in El Paso, Texas," *Feminismos en la antropología: nuevas propuestas críticas*, XI Congreso de Antropología (6), 191–206, www.ankulegi.org/.

102   one in four children: Erin Coulehan, "More Than 25 Percent of Children in El Paso Live in Poverty, Report Says," KTSM, August 31, 2021, www.ktsm.com/.

102    $17,000 less per year: Comparing 2022 data from the "El Paso County Profile," as compiled by the Texas Association of Counties, with the U.S. Bureau of Labor Statistics' "May 2022 National Occupational Employment and Wage Estimates."

103   third-largest center of garment manufacturing: Joel Zapata, "La Mujer Obrera of El Paso," Texas State Historical Association, November 22, 2013, www.tshaonline.org/.

103   Programa de Industrialización Fronteriza: Lawrence Douglas Taylor Hansen, "The Origins of the Maquila Industry in Mexico," *Comercio Exterior* 53, no. 11 (November 2003), www.revistas.bancomext.gob .mx/.

103   Bermúdez facility grew: Sklair, *Assembling for Development*, 111.

103   Connecticut Yankee: Jim Steinberg, "Making the Grade," *El Paso Herald-Post*, May 15, 1982.

103   El Paso needed a CEO: Randy Limbird, "They All Laughed When Mortgage Mogul Rogers Ran for Mayor," *El Paso Times*, June 14, 1989.

103   "tyrant-tycoon": Ibid.

104   "guayabera is a part of our culture": "Fun of His Own," *El Paso Times*, June 3, 1981.

104   "cut back neckties": "Mayor, Tie Ban Make National Big Time," *El Paso Times*, June 26, 1985.

104   city services suffered: David Crowder, "Era of Jonathan Rogers, the Stern Taskmaster, Closes," *El Paso Times*, June 11, 1989.

104   Rogers's counterpart in Juárez: Dick Reavis, "Showdown in Chihuahua," *Texas Monthly* (July 1986), www.texasmonthly.com/.

104   "What problems?": Tina Rosenberg, "The Prophets," *Texas Monthly* (February 1987), www.texasmonthly.com/.

105   "help both sides": Steve Brewer, "Peso Power Links Juarez to El Paso," Associated Press, May 17, 1981.

105   One hundred thousand Juárenses: Gay Young, "Gender Inequality and Industrial Development: The Household Connection," *Journal of Comparative Family Studies* 24, no. 1 (1993): 1–20, www.jstor.org/.

105   unemployment rate reached 13 percent: Sklair, *Assembling for Development*, 108–109.

105   close to a national celebrity: *Good Night America*, season 2, episode 18, "Willowbrook, Jay J. Armes," aired June 5, 1975, ABC.

105   "had a torpedo cupped": "Ysleta Child Loses Both Hands in Explosion of Railroad Torpedo," *El Paso Times*, May 15, 1946.

106   stabbed the incendiary: Gary Cartwright, "Is Jay J. Armes for Real?," *Texas Monthly* (January 1976).

106   feel-good follow-up story: Cecilia Napoles, "Ysleta Carrier Overcomes Handicap of Being Handless," *El Paso Times*, January 23, 1948.

106   "I broke into the movies": Armes, *Investigator*, 38–39.

106   call from Marlon Brando: Ibid., 15–27.

107   "you're not my type": Ibid., 25.

107   initial story about the rescue: "El Paso Investigator Finds Movie Stars' Son," Associated Press, March 14, 1972.

107   "loaded submachine gun": Cartwright, "Is Jay J. Armes for Real?"

107   "Twenty-Five Most Intriguing People": Alexis Madrigal, "5 Intriguing Things: Thursday, 2/13," *The Atlantic*, February 13, 2014, www.theatlantic.com/.

107   menagerie of safari animals: Ivor Davies, "The Private Eye of Texas Is Upon You," *Miami Herald*, June 25, 1975.

107   long list of impressive clients: Armes, *Investigator*, 95, 111, 133, 158.

107   "when a strange woman calls": Ibid., 95.

108   "charade played inside a nightmare": Ibid., 135.

108   Through a porthole window: Cartwright, "Is Jay J. Armes for Real?"

108   Gilded Age tobacco and insurance magnate: Geoffrey Hellman, "Rummaging Around with the Ryans," *New Yorker*, September 10, 1955.

108   cofounded Lockheed Aircraft: "Deaths: Thomas F. Ryan III," *Washington Post*, July 25, 1994, www.washingtonpost.com/.

108   Three Rivers Ranch: Cozzens, *Tres Ritos*, 93.

109   Ryan walked into his office: Armes, *Investigator*, 53–61.

110   "victim of the problems": Tim Novak, "'I'm a Victim,'" *El Paso Herald-Post*, April 12, 1985.

110   only 140 votes: Elisa Rocha, "Armes Says He'll Rattle Government," *El Paso Herald-Post*, May 29, 1989.

110   novelty clock on his desk: Interview with Bob Moore, June 14, 2022.

110   stand-up comedian and a cockroach: David Crowder, "City Representatives Debate, Trade Insults," *El Paso Times*, September 6, 1989.

110   Armes used the term "afro-engineered": Interview with Bob Moore, June 14, 2022.

111   garment industry in El Paso evaporated: D. Spener and R. Capps, "North American Free Trade and Changes in the Nativity of the Garment Industry Workforce in the United States," *International Journal of Urban and Regional Research* 25 (2001).

111   "They suddenly closed the factory": Hernández, "Dislocation and Globalization of the United States–Mexico Border," 191–206, www.ankulegi.org/.

111   Lee, Wrangler, and Levi's: Joel Zapata, "La Mujer Obrera of El Paso," Texas State Historical Association, November 22, 2013, www.tshaonline.org/.

NOTES

highest number in the United States: Spener and Capps, "North American Free Trade and Changes in the Nativity of the Garment Industry Workforce in the United States."

a third of El Paso's labor force: Comparing data from "Fort Bliss: Economic Impact on the Texas Economy, 2019," a report by the Texas Comptroller of Public Accounts, with the U.S. Bureau of Labor Statistics' "Economy at a Glance" data from 2019.

"not beauty-parlored well enough": Gilb, *Last Known Residence of Micky Acuña*, 14–15.

a plant capable of refining lead: Martin Donell Kohout, "ASARCO," Texas State Historical Association, November 1, 1994, www.tshaonline.org/; Timmons, *El Paso*, 176.

on a portion of company land: Martin Donell Kohout, "Smeltertown," Texas State Historical Association, July 1, 1995, www.tshaonline.org/.

San José de Cristo Rey: Polk, *El Paso's Mount Cristo Rey*; interview with Reuben Escandon, June 14, 2022.

inflamed by a sense of vanity: Bud Newman, "Urbici Soler as I Knew Him," *NOVA: The University of Texas at El Paso Magazine* (December 1972), www.issuu.com/.

"This spot is now consecrated": Bernice Zuniga and Terri Fout, "King on the Mountain," *Borderlands* 10 (1992), www.epcc.libguides.com/.

volunteers had organized a telethon: Interview with Ruben Escandon, June 14, 2022.

suffering permanent brain damage: Lauren Villagran, "Before Flint, Before East Chicago, There Was Smeltertown," National Resources Defense Council, November 29, 2016, www.nrdc.org/.

DEFIANCE

"I won't give up": Claytee White, "Oral History: Marzette Lewis," UNLV Special Collections and Archive, October 30, 2012, www.guides .library.unlv.edu/.7.

"that's how we survived": Myoung-ja Lee Kwon, "An Interview with Sook-ja Kim," UNLV Las Vegas Women in Gaming and Entertainment Oral History Project, 1997, www.special.library.unlv.edu/.

orientalist spectacular: Gerdes, *Sue Kim*, 69–75; Neil Genzlinger, "Dorothy Toy, 102, Half of Asian-American Dance Team, Dies," *New York Times*, August 1, 2019, www.nytimes.com/; "Obituary: Paul Wing," *San*

Francisco Chronicle, May 2, 1997, www.sfgate.com/; Chung, *The Chinese in Nevada*, 96.

122 "I don't think this is America": Lee Kwon, "An Interview with Sook-ja Kim."

122 "no buildings, no trees, only signs": Tronnes, *Literary Las Vegas*, 5.

122 "there is no place": Scott Brown, Venturi, and Izenour, *Learning from Las Vegas*, 18.

123 "a pretend city": Bégout, *Zeropolis*, 84, 18.

124 a small group of Mormons: Denton and Morris, *The Money and the Power*, 93.

124 boomtown nearest the Comstock Lode: "Virginia City Historic District, Nevada," National Park Service, June 5, 2018, www.nps.gov/.

124 fluff grass and red grama: Denton and Morris, *The Money and the Power*, 93; S. R. Abella and J. E. Craig, "Vegetation of Grassy Remnants in the Las Vegas Valley, Southern Nevada," *Desert Plants* 24 (2008): 16–23, www.digitalscholarship.unlv.edu/.

124 William Clark: "Las Vegas and the Railroad," UNLV Institute of Museum and Library Services, 2009, www.special.library.unlv.edu/.

125 "Conditions will be very little different": "Nevada Marks 90th Anniversary of Legal Gambling," Mob Museum Blog, March 19, 2021, www.themobmuseum.org/.

125 five thousand laborers: "Boulder City," *Nevada Experience*, KNPB, aired 1999, www.pbs.org/.

125 appealing destination: Denton and Morris, *The Money and the Power*, 94.

125 first crook to set up shop: Ibid., 98; "Las Vegas: An Unconventional History—Guy McAfee (1888–1966)," *American Experience*, PBS, November 14, 2005, www.pbs.org/.

126 Bugs and Lansky Mob: Denton and Morris, *The Money and the Power*, 49–58.

126 control of the gambling boats: Ibid., 50.

126 fees generated by that branch of the wire: Reid and Demaris, *Green Felt Jungle*, 15–17.

126 "goddamn biggest, fanciest gaming casino": Denton and Morris, *The Money and the Power*, 53.

127 "poor man's capital": Don Dedera, "Walter Reed Bimson: Arizona's Indispensable Man," *Arizona Highways* (April 1973).

127 massive, high-upside loans: Denton and Morris, *The Money and the Power*, 151–52.

127 Working all possible angles: Reid and Demaris, *Green Felt Jungle*, 20–25.

127 building a winner: Denton and Morris, *The Money and the Power*, 55–56.

127 rooms were not completed: Ibid., 53.

128 "Have we been introduced?": Reid and Demaris, *Green Felt Jungle*, 27.

128 "make little people feel big": Denton and Morris, *The Money and the Power*, 34.

128 bailed out by Bimson: Ibid., 59.

128 "We only kill each other": Reid and Demaris, *Green Felt Jungle*, 25.

128 Twenty minutes after Siegel's assassination: Ibid., 34–40.

129 Eight million visitors: Denton and Morris, *The Money and the Power*, 128.

129 Moulin Rouge had already closed: Kevin Cook, "The Vegas Hotspot That Broke All the Rules," *Smithsonian Magazine* (January 2013), www.smithsonianmag.com/.

129 debt-laden Royal Nevada: "Royal Nevada in U.S. Court," *Reno Gazette-Journal*, March 20, 1958.

129 "nowhere near the saturation point": Bill Becker, "Will Golden Bubble Burst?," *Reno Gazette-Journal*, January 14, 1956.

129 San Francisco rumrunner: Reid and Demaris, *Green Felt Jungle*, 73–79.

130 "three prim Korean girls": Gerdes, *Sue Kim*, 107–109.

130 Ed Sullivan was in town: Benjamin Han, "Transpacific: The Kim Sisters in Cold War America," *Pacific Historical Review* 87, no. 3 (2018): 473–98, www.jstor.org/.

130 launched them into genuine stardom: Gerdes, *Sue Kim*, 118–25.

130 unglamorous Robinson Apartments: Lee Kwon, "An Interview with Sook-ja Kim."

130 "No time to think!": Gerdes, *Sue Kim*, 136.

131 hang around the Stardust: Ibid., 140.

131 Elvis Presley came on to her: Ibid., 136.

131 "keep going forever": Ibid., 141.

131 Korean Kittens: "Korean Kittens in New Role at Thunderbird Hotel," *Santa Cruz Sentinel*, February 6, 1966.

131 staggering down the street in a bathrobe: Gerdes, *Sue Kim*, 187–97.

131 abruptly left for Los Angeles: Ibid., 226.

131 "working week-to-week": Ibid., 258–59.

132 from the Hilton to the Holiday Casino: Ibid., 269–79.

132 Dae Han Sisters: "Dae Han Sisters Are Marvey at Harveys," *San Francisco Examiner*, July 21, 1991.

132 "Maybe too comfortable": Keith Tuber, "Lounging with the Dae Han Sisters," *Orange Coast Magazine* (October 1987).

133 McCurdy cautioned visitors: Chase McCurdy, "Threads in Time," Marjorie Barrick Museum of Art, Las Vegas, 2020.

134 "There was so much hope": Clayee White, "Oral History: Anna Bailey," UNLV Special Collections and Archive, March 3, 1997, www.guides.library.unlv.edu/.

134 to Las Vegas from the Mississippi Delta: Orleck, *Storming Caesars Palace*, 37.

134 sharecroppers outside of Tallulah: Ibid., 11–36.

134 Black southerners who were heading west: "The Second Great Migra-

tion," *In Motion: The African American Migration Experience*, Schomburg Center for Research in Black Culture, 2005, www.inmotionaame.org/.

135 trailers, canvas tents, and wooden sheds: Hershwitzky, *West Las Vegas*.

135 cart around metal basins: Orleck, *Caesars Palace*, 47–48.

135 the city's Black population: Ibid., 37.

135 Jackson Avenue on the Westside: Claytee White, "Historic Black Vegas: Jackson Avenue Nightclubs in the 1940s," Las Vegas Black Image, April 8, 2022, lasvegasblackimage.com/; Pam Goertler, "A Westside Story: A Story of the Clubs and Casinos in Segregated Las Vegas," *Casino Chip and Token News* 21, no. 4 (2008), www.ccgtcc-ccn.com/.

135 Moulin Rouge got so hot: Orleck, *Caesars Palace*, 62; Cook, "The Vegas Hotspot That Broke All the Rules."

135 showgirls in colorful fringe dresses: *Life*, June 20, 1955.

136 failed to make payroll: "Moulin Rouge Casino Closed; Debts Blamed," *Nevada State Journal*, October 12, 1955.

136 the same treatment as the celebrities: Orleck, *Caesars Palace*, 63–65.

136 headboard of her hospital bed: Ibid., 71–73.

136 "He tried to be a good father": Ibid., 73.

137 signed up for a sewing class: Ibid., 100.

137 likened welfare mothers to "vultures": Ibid., 91.

137 welfare horror stories: Ibid., 94–95.

137 "work for their keep": Ibid., 128.

137 Albuquerque mother named María Avila: MacKenzie Elsasser and Tixier y Vigil, *Las Mujeres*, 149–50.

138 direct payments to farmers: Claude T. Coffman, "Target Prices, Deficiency Payments, and the Agriculture and Consumer Protection Act of 1973," *North Dakota Law Review* 50, no. 2, art. 6 (1973), www.commons .und.edu/.

138 slashing corporate tax rates: James Steele, "How Four Decades of Tax Cuts Fueled Inequality," Center for Public Integrity, November 22, 2022, www.publicintegrity.org/.

138 "finally started to come out": Orleck, *Caesars Palace*, 100.

138 "twenty-two percent completely ineligible": Ibid., 137–38.

138 Miller's random audits: Ibid., 133–39.

138 "This is *the* pocketbook": Ibid., 140.

139 "hit with a brick": Ibid., 152.

139 Hundreds of welfare families assembled: Ibid., 155–65.

141 "It's kind of silly, these amounts": Eli Hager, "The Cruel Failure of Welfare Reform in the Southwest," ProPublica, December 31, 2021, www .lasvegassun.com/.

141 "we are going to be here": Stephanie Overton, "Jackson Avenue Street Projects Breaks Ground, Project Set to Complete June 2023," KLAS, December 5, 2022, www.8newsnow.com/.

142 one in four chance it's in hospitality: "Las Vegas-Henderson-Paradise,

NV Economy at a Glance," U.S. Bureau of Labor Statistics, November 2023, www.bls.gov/.

142   Culinary Workers Local 226: Steven Greenhouse, "Labor Rolls On in Las Vegas, Hotel Union Is a National Model," *New York Times*, April 27, 1998, www.nytimes/com/.

143   "for the next hundred years": "Las Vegas Setting New Standards for Labor Movement," *Las Vegas Sun*, February 12, 1997, www.lasvegassun.com/.

143   raise pay and cover back wages: Robert Macy, "6½-Year Las Vegas Hotel Strike Ends," *Washington Post*, February 2, 1998, www.washingtonpost.com/.

143   "maids can own their homes": Lou Cannon, "Las Vegas's Service Industry Workers Hit the Jackpot with Union Contracts," *Washington Post*, November 30, 1997, www.washingtonpost.com/.

143   finding a cancerous growth: Claytee White, "An Interview with Hattie Canty," Boyer Las Vegas Early History Project, UNLV, 2007.

143   Culinary Workers endorsed Hillary Clinton: Jon Ralston, "UNITE HERE, Culinary Union Finally Endorses Hillary," KTNV, July 18, 2016, www.ktnv.com/.

144   Jesse Jackson and Rich Trumka: Interview with Ted Pappageorge, March 3, 2023.

144   half of nonunion workers: Heidi Shierholz, Celine McNicholas, Margaret Poydock, and Jennifer Sherer, "Workers Want Unions, but the Latest Data Point to Obstacles in Their Path," Economic Policy Institute, January 23, 2024, www.epi.org/.

145   upward of $12 billion every year: Average of total gaming revenue for Clark County from 2021 and 2022, as compiled by the Las Vegas Convention and Visitors Authority Research Center in "Las Vegas Historic Tourism Statistics (1970–2022)."

145   "may walk unmolested down the sidewalk": Hickey, *Air Guitar*, 19.

146   bingo parlors in upstate New York: Binkley, *Winner Takes All*, 4–7.

146   "The place itself was a spectacle": "Hotelier and businessman Steve Wynn talks about his love for Las Vegas, his resorts, and putting Picasso in a casino," *Charlie Rose*, PBS, aired July 20, 1997, www.charlierose.com/.

146   bought 3 percent of the Frontier Hotel: Fox, *In the Desert of Desire*, 16; Binkley, *Winner Takes All*, 49.

146   Detroit mob boss: "As Vegas Investigates Possible Crime Links," UPI, November 14, 1967.

146   Howard Hughes had just bought the Sands: Denton and Morris, *The Money and the Power*, 277.

146   a banker from Utah: Ibid., 154–67.

147   "quick, honest, and had an exceptional IQ": Nina Munk, "Steve Wynn's Biggest Gamble," *Vanity Fair* (June 2005), www.vanityfair.com/.

147   youngest person to ever run a casino: Binkley, *Winner Takes All*, 24; Jon

Christensen, "The Greening of Gambling's Golden Boy," *New York Times*, July 6, 1997, www.nytimes.com/.

147 "the advice of this man": John Katsilometes, "As Wynn Recalls, E. Parry Thomas Was 'the Guy,'" *Las Vegas Review-Journal*, September 8, 2016, www.reviewjournal.com/.

147 "my fifth son": Ed Koch, "E. Parry Thomas, Business Icon Who Helped Shape Las Vegas, Dies at 95," *Las Vegas Sun*, August 26, 2016, www.las vegassun.com/.

148 "black Donald Duck T-shirt": Binkley, *Winner Takes All*, 55.

148 happy to work the crowd: Ibid., 261–62.

148 swimming with his pets in a wetsuit: Ibid., 45.

149 went on an art-buying binge: Fox, *Desert of Desire*, 17; "Ruling: Public Must Have Free Access to Bellagio Art," *Las Vegas Sun*, August 19, 1998, www.lasvegassun.com/.

149 to underestimate the taste of the public: Binkley, *Winner Takes All*, 110.

149 Kirk Kerkorian: Ibid., 5–17.

149 would leap at a takeover offer: Andrew Pollack, "MGM Grand Makes $3.3 Billion Unsolicited Offer for Mirage," *New York Times*, February 2, 2000, www.nytimes.com/.

149 in a deal worth $6.4 billion: Robert Macy, "Stockholders, Nevada Regulators Approve $6.4 Billion Merger," *Las Vegas Sun*, May 31, 2000, www .lasvegassun.com/.

149 site of the old Desert Inn: Ibid., 203.

149 "a pattern of sexual misconduct": Alexandra Berzon, Chris Kirkham, Elizabeth Bernstein, and Kate O'Keeffe, "Dozens of People Recount Pattern of Sexual Misconduct by Las Vegas Mogul Steve Wynn," *Wall Street Journal*, January 27, 2018, www.wsj.com/.

149 netted him over $2 billion: Farah Master, "Steve Wynn Sells Stake in Company He Founded, Macau Casino Galaxy Buys In," Reuters, March 23, 2018, www.reuters.com/.

149 childhood friend D. Boone Wayson: "Wynn Resorts' Next Battle: Friends of Steve Who Remain on the Board," Bloomberg, April 26, 2018, www.news.bloomberglaw.com/.

150 "the only crime is getting caught": Thompson, *Fear and Loathing in Las Vegas*, 72.

151 49 million LED lights: "Everything You Need to Know about the Viva Vision Canopy Upgrade," Fremont Street Experience blog, July 27, 2023, www.vegasexperience.com/.

151 built from the ground up in downtown Vegas: Kiko Miyasato, "A Legend Is Born with Circa Resort & Casino in Las Vegas," *Las Vegas Magazine*, December 23, 2020, www.lasvegasmagazine.com/.

152 plus a mandatory $500 tip: Jay Caspian Kang, "What Would a Nation of Sports Gamblers Look Like?," *New Yorker*, October 21, 2022, www .newyorker.com/.

152   wore the same get-up: Molly O'Brien, "I'm a Dancing Card Dealer at a Vegas Casino. I Love My Job—Here's What It's Like," *Business Insider*, October 14, 2021, www.businessinsider.com/.

153   Bob Stupak's Glitter Gulch: Ed Koch and Mary Manning, "Bob Stupak, Builder of Stratosphere and Vegas World, Dies at 67," *Las Vegas Sun*, September 25, 2009, www.lasvegassun.com/.

153   slipped the apostrophe out of Caesars Palace: Bradley Martin, "The New Apostrophe Bar's Old School Tribute to Caesars," *Eater Las Vegas*, January 6, 2014, www.vegas.eater.com/.

153   nobody with juice: Reid and Demaris, *Green Felt Jungle*, 20.

153   home state of Michigan: Howard Stutz, "A Gambler at Heart: Derek Stevens Opening First All-New Resort in Four Decades in Downtown Las Vegas," *Nevada Independent*, October 25, 2020, www.thenevada independent.com/.

154   American Coney Island chili dog stand: Ron Sylvester, "Derek Stevens Tries to Bring a Bit of Detroit Downtown with the D," *Las Vegas Sun*, October 10, 2012, www.vegasinc.lavegassun.com/.

154   "Win Derek's Car": Jason Scavone, "D Man," *KNPR Desert Companion* (December 2016), www.knpr.org/.

154   introducing Derek Stevens to a national audience: Darren Rovell, "Man Could Win $1M if MSU Wins It All," ESPN, March 30, 2015, www .espn.com/; Joe Drape, "With 50–1 Bet, Casino Owner Sits Two Michigan State Wins from $1 Million," *New York Times*, April 4, 2015, www .nytimes.com/.

155   "a feeling of utter stupefaction": Bégout, *Zeropolis*, 46.

155   "overexcited atmosphere of the place": Ibid., 29.

156   "a sense of anomie": E. Hertz, T. Haywoode, and L. T. Reynolds, "The Sunbelt Syndrome: Radical Individualism at the End of the American Dream," *Humanity & Society* 14, no. 3 (1990): 257–79.

COMBUSTION

159   "The Southwest is a place": Charles Bowden, *Blue Desert*, 35.

159   house's rickety walls: Baca, *A Place to Stand*, 7–9.

159   "white-skinned, blue-eyed": Ibid., 14.

160   "be on your best behavior": Ibid., 15.

160   "back in two weeks": Daniel Glick, dir., *A Place to Stand*, Catamount Films, 1994.

160   link of my true blood: Baca, "Silver Water Tower," in *Singing at the Gates*, 82.

160   Catholic orphanage: Gregory Hicks and Rolf M. Evenson, "Historic American Building Survey: St. Anthony's Orphanage," National Park Service, www.loc.gov/.

160   beaten by the nuns: Glick, *A Place to Stand*.

161   "I was a deviant": Baca, *A Place to Stand*, 4.

161   kill the Indian: Richard Henry Pratt, "The Advantages of Mingling Indians with Whites," speech, National Conference of Charities and Correction, Denver, CO, June 23–29, 1892, Dickinson College Carlisle Indian School Digital Resource Center.

161   "gradual uplifting of the race": Lillie G. McKinney, "History of the Albuquerque Indian School," *New Mexico Historical Review* 20, no. 2 (1945), www.digitalrepository.unm.edu/.

161   boot camp–like environment: Howard Ellis, "From the Battle in the Classroom to the Battle for the Classroom," *American Indian Quarterly* 11, no. 3 (Summer 1987): 255–64, www.jstor.org/.

161   thirty students had been buried: Kalen Goodluck, "The Children at Rest in 4-H Park," *High Country News*, March 29, 2022, www.hcn.org; Jonathan Sims, "Indian School Graves Rediscovered Under City Park," *Paper*, July 21, 2021, www.abq.news/.

161   man alleged being sexually abused: Colleen Heild, "Lawsuit Alleges St. Anthony Home for Boys Was Rife with Abuse," *Albuquerque Journal*, August 6, 2020, www.bishop-accountability.org/; "Two New Sexual Abuse Claims Against St. Anthony's Orphanage Nun," KRQE, August 25, 2020, www.krqe.com/.

162   "a gardener's shack": Baca, *A Place to Stand*, 31.

162   "alone in the world": Ibid., 35.

162   socially censured whenever I was in public: Ibid., 4.

163   "This land was Mexican once": Anzaldúa, *Borderlands*, 25.

163   census taker: Simón Ventura Trujillo, "Forgotten Pueblos: *La Alianza Federal de Mercedes* and the Cultural Politics of Indo-Hispano," PhD diss., University of Washington, 2013.

164   marked the 115th anniversary: Martínez and Longeaux y Vásque, *Viva la Raza!*, 156–59.

164   Santa Fe Ring: Philip J. Rasch, "The People of the Territory of New Mexico vs. the Santa Fe Ring," *New Mexico Historical Review* 47, no. 2 (2021), www.digitalrepository.unm.edu/.

164   Tierra Amarilla: Charles Butler, "History of the Tierra Amarilla Grant, Northern New Mexico," New Mexico Geological Society, 1977, www.nmgs.nmt.edu/.

164   more than 2 million acres: Victor Westphall, "Thomas Benton Catron: A Historical Defense," *New Mexico Historical Review* 63, no. 1 (1988), www.digitalrepository.unm.edu/.

165   Spanish-language radio: Martínez and Longeaux y Vásque, *Viva la Raza!*, 159.

165   National Indian Youth Council: Shreve, *Red Power Rising*.

165   Tijerina shifted the argument: Josue David Cisneros, "Reclaiming the

Rhetoric of Reies López Tijerina: Border Identity and Agency in 'The Land Grant Question,'" *Communication Quarterly* 60, no. 5 (2012): 561–87.

166 The municipal model: Luckingham, *Urban Southwest*, 86; Needham, *Power Lines*, 105.

166 Phoenix's manufacturing output: Luckingham, *Urban Southwest*, 96.

166 as far as agriculture went: Ibid., 99.

166 "spawned in the minds of those who are idle": Ibid., 119.

166 "my sense of outrage": Elsasser, MacKenzie, and Tixier y Vigil, *Las Mujeres*, 87–89.

166 "slum clearance" effort: Robert Fairbanks, "The Failure of Urban Renewal in the Southwest: From City Needs to Individual Rights," *Western Historical Quarterly* 37, no. 3 (Autumn 2006): 303–25, www.jstor.org/.

167 Tijeras Urban Renewal Project: Max Boruff, "A Changing Vision: Albuquerque's Urban Renewal," UNM School of Architecture and Planning, 2015, www.albuquerquemodernism.unm.edu/.

167 appealed to the federal government: Fairbanks, "The Failure of Urban Renewal in the Southwest."

167 new police headquarters: Mickey Reilly, "Downtown Renewal End in Sight," *Albuquerque Tribune*, March 12, 1974.

167 "no sense of participation": Baca, "¿De Quiéncentennial? Puro Pedo . . . ," in *Working in the Dark*, 95.

168 Bobby Garcia: Antonio Moreno, "Remembering Rito Canales and Antonio Cordova of the Black Berets of Albuquerque," Siglo de Lucha, March 3, 2017, www.siglodelucha.wordpress.com/.

168 informers for the Metro Squad: Kalen Goodluck, "New Mexico's Thin Blurred Line," *High Country News*, July 20, 2020, www.hcn.org/.

168 federal civil rights investigation: Sterling F. Black, chairman, "The Struggle for Justice and Redress in Northern New Mexico," New Mexico Advisory Committee to the United States Commission on Civil Rights, 1974.

168 soldiers affixed bayonets: Rick Nathanson, "Bayoneting Victims Recall 50th Anniversary of Bloodshed at UNM," Albuquerque Journal, May 9, 2020; Christopher Beaudet, "Eleven Bayonetted and 131 Arrested at Student Union Building," University of New Mexico, 2021.

168 stabbing ten students and a cameraman: Martin Waldron, "Guardsmen Defended by Chief Over Stabbings in New Mexico," *The New York Times*, June 7, 1970.

168 "Let's get the pigs!": Mike Padget, "Reports of Incidents Conflict," *Albuquerque Journal*, June 16, 1971.

168 "the wildest night in Albuquerque's history": Tomas Martinez and Mike Padget, "Nine Shot as Police Battle Mob," *Albuquerque Tribune*, June 14, 1971.

169 march to the city hall: This scene has been recreated using a variety of

contemporaneous news reports from both *The Albuquerque Tribune* and the *Albuquerque Journal* from June 14 to 16, 1971, namely: Martinez and Padget, "Nine Shot as Police Battle Mob; Ed Mahr and Tomas Martinez, "Troops Help Police, Move in After Mob Marches," *Journal*, June 15, 1971; Ralph Looney and *Tribune* Staff, "2nd Night of Terror for Albq! Roving Bands Loot and Burn; Jail 283," *Tribune*, June 15, 1971; Harry Moskos, "Police Chief Byrd's First Interview on the Riots in Albq," *Tribune*, June 6, 1971; Mike Padget, "Reports of Incidents Conflict," *Journal*, June 16, 1971.

170   remains of a school administration building: "Riot at Roosevelt Park— 50 Year Anniversary," KUNM, June 11, 2021, www.kunm.org/.

170   "not alleged but factual": Bruce Herron and Charles Wood, "Mondragon Denies Talk in Park Incited Rioting," *Albuquerque Tribune*, June 16, 1971.

170   "alleged police brutality": "Bumbling Riot Control," *Albuquerque Journal*, June 16, 1971.

170   hands were handcuffed: Black, "The Struggle for Justice and Redress in Northern New Mexico."

170   construction site on the Westside: *Cordova v. Larson*, Court of Appeals of New Mexico, no. 23,846, May 19, 2004, www.caselaw.findlaw.com/.

170   "When we burned": Baca, "Martín," in *Martín and Meditations on the South Valley*, 6.

171   "evidence of the good life": Baca, *A Place to Stand*, 43–44.

171   "I lived in the streets": Baca, "Martín," 20.

172   "Spanish origin": Gómez, *Inventing Latinos*, 152–60.

172   five-year confinement: Baca, *A Place to Stand*, 36–42, 71–88.

173   "I can't describe how words electrified me": Ibid., 185.

173   served time in prison: Thomas Bonczar, "Prevalence of Imprisonment in the U.S. Population, 1974–2001," U.S. Department of Justice, Bureau of Justice Statistics, August 2003, www.bjs.ojp.gov/.

174   Lincoln Continental: Baca, *A Place to Stand*, 259–62.

174   notorious local law enforcement agencies: Rachel Aviv, "Your Son is Deceased," *New Yorker*, January 26, 2015, www.newyorker.com/.

174   nonwhite residents still struggle: "An Equity Profile of Albuquerque," W. K. Kellogg Foundation, June 2018, www.nationalequityatlas.org/.

175   bookseller's convention in Orange County: Beth Ann Krier, "Baca: A Poet Emerges from Prison of His Past," *Los Angeles Times*, February 15, 1989, www.latimes.com/.

175   "open more doors with *Hispanic*": Gomez, *Becoming Latino*, 141.

175   "My Dog Barks": Baca, in *Singing at the Gates*, 153.

VIABILITY

181   "The 'street' I lived on": Martínez, *Desert America*, 2.

182   Home Owners' Loan Corporation: Richard Rothstein, *Color of Law: A*

Forgotten History of How Our Government Segregated America (New York: W. W. Norton & Company, 2017), 63–67.

182   South Phoenix was redlined by HOLC: Katie Gentry and Alison Cook-Davis, "A Brief History of Housing Policy and Discrimination in Arizona," ASU Morrison Institute for Public Policy, November 2021, www.morrisoninstitute.asu.edu/; Elise Miller, "Connecting the Dots Between Redlining and Heat Resilience in Phoenix," ASU Knowledge Exchange for Resilience, June 22, 2021, www.news.asu.edu/.

182   Greenwood District of Tulsa: Ben Fenwick, "The Massacre That Destroyed Tulsa's 'Black Wall Street,'" *New York Times*, July 13, 2020, www.nytimes.com/.

182   secreting himself into a train car: Whitaker, *Race Work*, 36–37.

182   "I wanted to be a pilot": Ibid., 44.

183   no running water or bathrooms: Ibid., 92.

183   "public utilities inadequate": Needham, *Power Lines*, 85.

183   "Arizona's Finest Negro Mortuary": Whitaker, *Race Work*, 97–104.

183   whites-only communities: Ibid., 105.

184   first iteration of the Heard Museum: "Heard Museum History in the 1929 Gallery," Heard Museum, Phoenix, 2024.

184   Lloyd Larkin and George Peter: George and Conrad, *Phoenix's Greater Encanto-Palmcroft Neighborhood*, 19, 22, 28.

184   2040 Encanto Drive: Tricia Amato, "1928: The Indian House," *This Old House Phoenix*, July 31, 2018, www.thisoldhousephoenix.com/.

184   had her own real estate license: Whitaker, *Race Work*, 108–109.

185   refused to clean up: Ibid., 110.

185   Black homeownership: Catherine Reagor and Megan Taros, "'Institutionalized Racism of the Past': Discriminatory Housing Practices Resound in South Phoenix Today," *Arizona Republic*, April 18, 2022, www.azcentral.com/.

185   using a GI loan: Needham, *Power Lines*, 67–69.

185   numerous offers to buy it: Jeffrey Horst, "The Maryvale Project," ASU Cronkite School Innovation Lab, June 2022, www.cronkitenews.azpbs.org/.

185   reinvesting his profits: "Builder Noted for Success," *Arizona Republic*, February 5, 1956.

185   "they made the loan": Pam Stevenson, "Arizona Historymakers: John F. Long," Historical League, Inc., August 29, 2000, www.historicalleague.org/.

185   annexed by the city of Phoenix: Don Dedera, "Maryvale Prepares 'One Big Blossom,'" *Arizona Republic*, March 26, 1964.

185   88 of the new residents: Horst, "Maryvale Project."

186   seized land in the barrios: "Gangs: Violent Youths Reflect Deep Change on City's West Side," *Arizona Republic*, October 28, 1979.

186   gravestone for the Golden Gate: Eduardo Barraza, "A Vanished Phoenix

Barrio: Visions of Life on 16th Street," *Barriozona*, October 15, 2004, www.barriozona.com/.

186 neighborhood became a destination: Horst, "Maryvale Project."

186 declared a Superfund site: Ross, *Bird on Fire*, 128–29.

186 "people lose their sense of community": "Gangs: Violent Youths Reflect Deep Change on City's West Side," *Arizona Republic*, October 28, 1979.

187 educational attainment was also declining: "West Phoenix Revitalization Area Economic Background Plan," ESI Corporation, July 1, 2008, www.phoenix.gov/.

187 owner-occupied: "2010 Census Summary File 1: Maryvale Village," City of Phoenix Planning & Development Research Team, February 16, 2012, www.phoenix.gov/.

187 number of miles traveled: From a comparison of "The Phoenix Metro Area," Maricopa Association of Governments, January 2020, azmag.gov/, and "Hits and Misses: Fast Growth in Metropolitan Phoenix," ASU Morrison Institute for Public Policy, September 2000, www.morrisoninstitute.asu.edu/.

187 exhaust produced in Phoenix: Lauren Gilger, "Phoenix Area Vehicle Pollution Increased 291% Since 1990," KJZZ, November 1, 2019, www.kjzz.org/.

187 worst air pollution in the country: "Most Polluted Cities," American Lung Association, 2023, www.lung.org/.

187 wealthy enclaves in the same cities: Ian James, "Low-Income and Latino Neighborhoods Endure More Extreme Heat in the Southwest, Study Shows," *Arizona Republic*, March 15, 2021, www.azcentral.com/.

187 drafty manufactured homes: Caroline Tracey, "'You're Living in a Tin Can,'" *High Country News*, September 28, 2022, www.hcn.org/.

188 supplies air conditioning units: "Household Energy Use in Arizona," Residential Energy Consumption Survey, U.S. Energy Information Administration, 2009, www.eia.gov/.

188 generated by renewable sources: "Arizona State Energy Profile," U.S. Energy Information Administration, May 18, 2023, www.eia.gov/.

188 angular spaces of the Sonoran Desert: Guided tour of Taliesin West on February 14, 2020.

189 "prescription for a modern house": Wright, *In the Realm of Ideas*, 44.

189 "complementary to its natural environment": Curtis, *Modern Architecture Since 1900*, 312.

190 personal appeal to President Harry Truman: Melissa Lawford, "Phoenix Development Causes Saturation in the Desert," *Financial Times*, March 15, 2019, www.ft.com/.

190 plants burning coal: Needham, *Power Lines*; Powell, *Landscapes of Power*.

190 pair met in West Texas: Interview with Hayes and Robles, March 30, 2023.

191 Zaguáns: Vint and Neumann, *Southwest Housing Traditions*, 53–56.

192 "expression of ancient values": Canizaro, *Architectural Regionalism*, 188–93.

192 seventeen-point plan: Katie Weeks, "The Green Phoenix Plan Aims to Make the Desert City the Greenest City in America," *Architect Magazine*, November 9, 2009, www.architectmagazine.com/.

192 expand that light-rail: Danny Shapiro, "Mayor Kate Gallego Boosts Light Rail During State of Phoenix Address," KTAR, June 14, 2019, www.ktar.com/.

192 "the most sustainable desert city": "Climate Action Plan Framework," City of Phoenix, November 2020.

192 per capita use in the city: Fernanda Santos, "An Arid Arizona City Manages Its Thirst," *New York Times*, June 16, 2013, www.nytimes.com/; Wayne Solley, "Estimates of Water Use in the Western United States in 1990 and Water-Use Trends 1960–90," U.S. Geological Survey, August 1997, www.nrm.dfg.ca.gov/.

193 ad in *The Arizona Republic:* "Water Rates Change Today," *Arizona Republic*, June 1, 1990.

193 first families sent by Brigham Young: Interview with Kathryn Sorensen, April 5, 2023.

193 water consumption in July: "Water Resource Plan," City of Phoenix, 2021, 74.

194 consumption actually went up: Rachel Ramirez, "California Is in a Water Crisis, Yet Usage Is Way Up. Officials Are Focused on the Wrong Problem, Advocates Say," CNN, May 15, 2022, www.cnn.com/.

194 Exurban growth doubles the impact: H. Heidari, M. Arabi, T. Warziniack, and S. Sharvelle, "Effects of Urban Development Patterns on Municipal Water Shortage," *Frontiers in Water* 3 (2021).

195 growing an acre of cotton: Abrahm Lustgarten and Naveena Sadasivam, "Killing the Colorado: How Federal Dollars Are Financing the Water Crisis in the West," ProPublica, May 27, 2015, www.projects.propublica.org/.

195 Roosevelt Row scene: Lynn Trimble, "Who Deserves Credit for Making Roosevelt Row a Thriving Part of Downtown Phoenix?," *Phoenix New Times*, April 4, 2016, www.phoenixnewtimes.com/.

195 Sappho's Fist: "Blabbing with Sappho's Fist," Dead Angel, 1996, www.xeenation.com/fist/interview.html/.

195 entirely nude cast: Kyle Lawson, "Planet Earth Theatre," *Phoenix Theater: An Eccentric History*, 2016, www.phoenixtheaterhistory.com/.

195 all-out urban festival: Ross, *Bird on Fire*, 93–94.

196 need for shelter still outpaces demand: Belinda Luscombe, "Why Phoenix—of All Places—Has the Fastest Growing Home Prices in the U.S.," *Time*, May 3, 2022, www.time.com/.

196 average rent of those units: Catherine Reagor and Jessica Boehm,

"Downtown Phoenix's Growing Popularity Is Pricing Out Many Residents," *Arizona Republic*, October 25, 2018, www.azcentral.com/.

196 subsidized housing voucher: "2020–2024 Consolidated Plan," City of Phoenix, June 5, 2020, 49.

196 the Zone: Eli Saslow, "A Sandwich Shop, a Tent City, and an American Crisis," *New York Times*, March 19, 2023, www.nytimes.com/.

196 homeless population of Maricopa County: "2023 Point-in-Time (PIT) Count Report," Maricopa County Association of Governments, January 23, 2023, www.azmag.gov/.

196 known as the "SOS lot": Helen Rummel, "Phoenix's Campground Now Open for People Who Are Experiencing Homelessness," *Arizona Republic*, November 13, 2023, www.azcentral.com/.

197 undeveloped land within Phoenix: "Housing Phoenix: Creating a Stronger and More Vibrant Phoenix Through Increased Housing Options for All," City of Phoenix, July 20, 2020, www.phoenix.gov/.

197 land could provide homes: Calculation based on 2020 census finding that Phoenix's population per square mile is 3,104.5 and Los Angeles's population per square mile is 8,304.2.

197 dubbed the "village core": "Maryvale Village Core Urban Design," City of Phoenix Planning and Development Department, 2011, www.phoenix .gov/.

197 killed by the heat: "2022 Heat Deaths Report," Maricopa County Department of Public Health, Division of Epidemiology and Informatics, June 2023, www.maricopa.gov/; Jessica Boehm, "Metro Phoenix Recorded 600+ Heat Deaths Last Year, but Optimistic for Summer 2024," Axios Phoenix, January 22, 2024, www.axios.com/.

197 heat-related deaths nationwide: Phillip Reese, "Heat-Related Deaths Are Up, and Not Just Because It's Getting Hotter," KFF Health News, September 8, 2023, www.kffhealthnews.org/.

197 pervasive homelessness crisis: "State of Homelessness: 2023 Edition," National Alliance to End Homelessness, 2023, www.endhomelessness .org/.

198 "expressions of the socio-political reality": Peter Zec, "Interview with Dieter Rams," in *Red Dot Design Yearbook* 2013/2014 (July 2013).

MUTUALISM

201 "And the dust that / confuses countries": Rosa Alcalá, "Inflection," in *The Lust of Unsentimental Waters* (Bristol: Shearsman Books, 2012), 17.

201 began their long journey: This scene was reconstructed using the series of reports that Louis Sahagún published in *The Tucson Citizen* between July 7 and 12, 1980: "Desert Heat Kills 13 Aliens: Smugglers Abandon 30 El Salvadorans," July 7; "Smugglers May Have Been Victims of Their

Own Errors," July 9; "4 Salvadoran Desert Victims May Have Been Strangled," July 10; "Her 3 Daughters Died in the Desert," July 10; and "Ten Salvadorans Released to Court-Appointed Sponsors," July 12.

204 Tucson's Southside Presbyterian Church: Ahead of his retirement in 2005, John Fife gave a series of four sermons detailing the history of Southside Presbyterian Church. Those sermons were recorded by a parishioner and uploaded to the Bandcamp page of OneWind Productions in 2019 as "The Oral History of Southside." The quotes that follow are largely taken from the second and third sermons, which cover the history of the church between 1970 and 1992.

204 "tiny country of titanic sorrows": Lovato, *Unforgetting*, xvii.

204 When *The Tucson Citizen* profiled John Fife: Tony Davis, "Minister in Blue Jeans: He Helps Tucson's Poor Find Way," *Tucson Citizen*, December 27, 1979.

205 declared themselves as sanctuaries: Ibid.

205 arrested an activist in South Texas: "Sanctuary Movement Leader Is Charged," Associated Press, December 1, 1984.

206 two-mile convoy: John Fife, "History of Southside 1981–1992," sermon at Southside Presbyterian Church, Tucson, 2005.

206 "alien smuggling": Mark Turner, Ernie Heltsley, and James H. Maish, "Indictments Target Sanctuary Members," *Arizona Daily Star*, January 15, 1985.

206 given probation by a judge: Daniel Browning, "Judge Criticizes INS, Sanctuary Movement After Giving Probation to Last 3 Defendants," *Arizona Daily Star*, July 3, 1986.

206 Sandinista government in Nicaragua: Paul Glickman, "Lawmakers Renew Push to Grant Asylum to Refugees," *Christian Science Monitor*, January 27, 1987, www.csmonitor.com/.

206 Temporary Protected Status: Eva Segerblom, "Temporary Protected Status: An Immigration Statute That Redefines Traditional Notions of Status and Temporariness," *Nevada Law Journal* 7, no. 2, art. 19 (2007), www.scholars.law.unlv.edu/.

206 arrests in the region south of Tucson: Kristina Davis, "Operation Gatekeeper at 25: Look Back at the Turning Point That Transformed the Border," *Los Angeles Times*, September 30, 2019, www.latimes.com/.

207 across the Arizona borderlands: Gabrielle Fimbres, "Still to Strive for Justice: Social Activist John Fife Will Retire Soon," *Tucson Citizen*, July 2, 2005.

207 "systematically herded onto death trails": C. T. Revere, "Humanitarians Call for Changes: Group Seeks Water Stations, Border Policy Shift," *Tucson Citizen*, May 31, 2001.

207 No More Deaths / No Más Muertes: Gutiérrez, "Do Migrants Dream of Blue Barrels?," in *Brown Neon*; Martínez, *Desert America*, 157–203.

207 drain jugs of water: Rory Carroll, "US Border Patrol Routinely Sabo-

tages Water Left for Migrants, Report Says," *Guardian*, January 17, 2018, www.theguardian.com/.

207 "ransacked and stripped": Cantú, *The Line Becomes a River*, 33–34.

207 Border Patrol's budget expanded: Comparing data from "The Cost of Immigration Enforcement and Border Security," American Immigration Council, January 2017, nnirr.org/, and "U.S. Customs and Border Protection Budget Overview: Fiscal Year 2023," Department of Homeland Security, March 2022, www.dhs.gov/.

207 ten times as many human remains: Curt Pendergrast and Alex Devoid, "Migrant Deaths: A Crisis Deepens in the Desert: Migrants Perishing in Greater Numbers," *Arizona Daily Star*, December 30, 2021, www .tucson.com/.

209 target of misogynist violence: Alicia Gaspar de Alba, "The Maquiladora Murders, or, Who Is Killing the Women of Juárez, Mexico?," UCLA Chicano Studies Research Center, August 2003, www.chicano.ucla.edu/.

209 200 officially recorded murders: Bowden, *Murder City*, 110.

210 Joint Chihuahua Operation: "Militarization Is Not the Solution," Centro de Derechos Humanos de las Mujeres, October 12, 2022, www.cedehm .org.mx/.

210 over ten thousand Juárenses had been slain: René Kladzyk, "Homicides in Juárez Reach Three-Year Low Amid Increased Attention to Femicides," El Paso Matters, February 3, 2022, www.elpasomatters.org/.

210 Iraq and Afghanistan: "Costs of War," Brown University Watson Institute for International & Public Affairs, July 2021, www.watson.brown .edu/.

210 federal troops withdrew from the city: William Booth, "In Mexico's Murder City, the War Appears Over," *Washington Post*, August 20, 2012, www.washingtonpost.com/.

210 mostly avoided El Paso del Norte: Martínez, *Beast*, 239–53.

210 fifty-two border obelisks: Alvarez, *Border Land, Border Water*, 17–52.

211 Mexican officials have responded: Interview with Enrique Valenzuela, June 15, 2022.

211 a policy called "metering": "Fact Sheet: Metering and Asylum Turnbacks," American Immigration Council, March 8, 2021, www.american immigrationcouncil.org/.

212 "Remain in Mexico" policy: Kirstjen Nielsen, "Policy Guidance for Implementation of the Migrant Protection Protocols," Department of Homeland Security, January 24, 2019, www.dhs.gov/; "Q&A: Trump Administration's 'Remain in Mexico' Program," Human Rights Watch, January 29, 2020, www.hrw.org/.

212 forced to reinstate MPP: Uriel García, "Judge Blocks Biden Administration from Lifting Public Health Order Used to Quickly Expel Migrants," *Texas Tribune*, May 20, 2022, www.texastribune.org/.

214 ready to be released: Nick Miroff, "Ruben Garcia Has Sheltered Migrants

in El Paso For over Four Decades. With a Crush at the Border, He Now Needs a Warehouse," *Washington Post*, April 1, 2019, www.washington post.com/.

214   dropped off downtown: Robert Moore, Ramon Bracamontes, and Corrie Boudreaux "Border Agents Once Again Release Migrants in Downtown El Paso," El Paso Matters, May 15, 2022, www.elpasomatters .org/.

214   influx of refugees from Venezuela: Uriel García, "El Paso Scrambles to Move Migrants Off the Streets and Gives Them Free Bus Rides as Shelters Reach Capacity," *Texas Tribune*, September 20, 2022, www.texas tribune.org/.

214   legally obliged to allow those migrants: Kimberly Atkins Stohr and John Ringer, "What's Behind the Recent Increase of Venezuelan Migrants in Texas," WBUR, September 26, 2022, www.wbur.org/.

214   Temporary Protected Status designation to Venezuelans: Regina Garcia Cano and Elliot Spagat, "With Temporary Status for Venezuelans, the Biden Administration Turns to a Familiar Tool," Associated Press, September 23, 2023, www.apnews.com/.

214   arrived at the southern border: David Shortell and Julie Turkewitz, "Venezuelans Who Left Everything Behind Are Stuck South of U.S. Border," *New York Times*, October 24, 2022, www.nytimes.com/.

214   Operation Lone Star: Cindy Ramirez, "El Paso Buses Hundreds of Migrants to New York City; More Released to the Streets," El Paso Matters, September 12, 2022, www.elpasomatters.org/; "Operation Lone Star Takes Historic Action Throughout 2022," Office of the Texas Governor Greg Abbott, December 29, 2022, www.gov.texas.gov/.

214   refugees had been sent to New York: J. David Goodman, "Texas Has Bused 50,000 Migrants. Now It Wants to Arrest Them Instead," *New York Times*, October 18, 2023, www.nytimes.com/.

215   number of migrant apprehensions spike: Comparing numbers from "Southwest Border Sectors Total Illegal Alien Apprehensions by Fiscal Year," U.S. Border Patrol, March 2019, www.cbp.gov/; and Julia Ainsley, "Migrant Border Crossings in Fiscal Year 2022 Topped 2.76 Million, Breaking Previous Record," NBC News, October 22, 2022, www.nbc news.com/.

215   "facilitating illegal entry": Rachel Monroe, "El Paso's Saint of the Border Negotiates a New Reality," *New Yorker*, February 23, 2024, www.new yorker.com/.

215   failed bid to shut his organization: Robert Moore, "El Paso Judge Blocks Texas AG Ken Paxton's Effort to Close Annunciation House," El Paso Matters, March 11, 2024, www.elpasomatters.org/.

215   10 percent of the American population moves: William Frey, "Americans' Local Migration Reached a Historic Low in 2022, but Long-

Distance Moves Picked Up," Brookings Institution, February 2, 2023, www.brookings.edu/.

216 order barring undocumented immigrants: "Fact Sheet: President Biden Announces New Actions to Secure the Border," White House Communications Office, June 4, 2024.

216 fire broke out: James Fredrik, "A Survivor Recalls Horrors of Mexico's Migrant Center Fire That Almost Killed Him," NPR, October 14, 2023, www.npr.org/.

216 "toward the train tracks": Martínez, *Beast*, 269.

217 "not running from hunger": Ibid., 272–73.

217 elsewhere around the world: Ashley Wu, "Why Illegal Border Crossings Are at Sustained Highs," *New York Times*, October 29, 2023, www.nytimes.com/.

217 disrupt the region's agriculture: Abrahm Lustgarten, "The Great Climate Migration," *New York Times Magazine*, July 23, 2020, www.nytimes.com/.

218 Rio Grande Rectification Project: Alvarez, *Border Land, Border Water*, 101–106.

219 "Third World grates against the first": Anzaldúa, *Borderlands*, 25.

219 Anglos expected to become a national minority: William Frey, "The US Will Become 'Minority White' in 2045, Census Projects," Brookings Institution, March 14, 2018, www.brookings.edu/.

219 majority-minority future is already here: Contemporary demographic data is culled from the 2020 Census.

219 Latin America and Asia: Abby Budiman, Christine Tamir, Lauren Mora, and Luis Noe-Bustamante, "Facts on U.S. Immigrants, 2018," Pew Research Center, August 20, 2020, www.pewresearch.org/.

219 "two or more races": Nicholas Jones, Rachel Marks, Roberto Ramirez, and Merarys Ríos-Vargas, "Improved Race and Ethnicity Measures Reveal U.S. Population Is Much More Multiracial," U.S. Census Bureau, August 12, 2021, www.census.gov/.

220 water access is less secure: Katie Meehan, Jason R. Jurjevich, Nicholas M. J. W. Chun, and Justin Sherrill, "Geographies of Insecure Water Access and the Housing–Water Nexus in US Cities," *PNAS* 117, no. 46, November 17, 2020, www.pnas.org/.

220 heat waves more intense: Nambi Ndugga and Samantha Artiga, "Continued Rises in Extreme Heat and Implications for Health Disparities," KFF, August 24, 2023, www.kff.org/.

220 more than forty family members: Myoung-ja Lee Kwon, "An Interview with Sook-ja Kim," UNLV Las Vegas Women in Gaming and Entertainment Oral History Project, 1997, www.special.library.unlv.edu/.

221 Asian Community Development Council of Nevada: Interview with Vida Lin, February 1, 2023.

221 Tucson bakery called Barrio Bread: Interview with Don Guerra, August 26, 2022.

223 chapalote corn: Gary Nabhan, "Chapalote Corn—The Oldest Corn in North America Pops Back Up," *Heirloom Gardener* (Winter 2012–13), www.garynabhan.com/.

223 white Sonora wheat: "White Sonora Wheat," Native Seeds Search, 2024, www.nativeseeds.org/.

225 four-hundredth anniversary of the conquistador: Alison Fields, "New Mexico's Cuarto Centenario: History in Visual Dialogue," *Public Historian* 33, no. 1 (Winter 2011): 44–72, www.digitalrepository.unm.edu/.

225 "someone who would be a token": Katy June-Friesen, "Recasting New Mexico History," *Weekly Alibi*, October 20, 2005, www.alibi.com/.

226 the right foot of the bronze figure: Simon Romero, "Statue's Stolen Foot Reflects Divisions Over Symbols of Conquest," *New York Times*, September 30, 2017, www.nytimes.com/.

226 Naranjo Morse told an interviewer: Harlan McKosato, "Native America Calling," Corporation for Public Podcasting, aired April 9, 2004.

227 firing at the crowd: Simon Romero, "Man Is Shot at Protest Over Statue of New Mexico's Conquistador," *New York Times*, June 15, 2020, www.nytimes.com/.

227 defending his work: Sonny Rivera, "Learn from Art, Don't Just Destroy It," *Albuquerque Journal*, June 28, 2020, www.abqjournal.com/.

228 more conciliatory tone: Deb Haaland, "In Troubled Times, We Can Celebrate Cultures of NM," *Albuquerque Journal*, June 28, 2020, www.abqjournal.com/.

228 her life story: Marco Della Cava and Deborah Barfield Berry, "'Our Ancestors' Dreams Come True': Deb Haaland Becomes the Nation's Most Powerful Native American Leader," *USA Today*, March 15, 2021, www.usatoday.com/.

228 was taken down: Joni Auden Land, "Oñate Statue in Alcalde Removed for Safekeeping," *Albuquerque Journal*, June 15, 2020, www.abqjournal.com/.

229 county's main offices in Española: "Oñate Statue to Return to Rio Arriba," *Rio Grande Sun*, September 20, 2023, www.riograndesun.com/.

229 would be postponed: "Return of Controversial Juan de Oñate Statue Postponed," KOB, September 28, 2023, www.kob.com/.

229 activists stood and prayed: Scene reconstructed using posts from the Red Nation Movement's Instagram page from September 28 through October 1, 2023, www.instagram.com/.

230 not move forward with a reinstallation: Megan Gleason and Shaun Griswold, "Man in Police Custody for Shooting Person During Prayer to Oppose Oñate," Source New Mexico, September 28, 2023, www.sourcenm.com/.

SPRAWL

235   "The deeds and papers don't mean anything": Silko, *Ceremony*, 118.

235   numerous neighborhoods in Cadence: Eli Segall, "Q+A: Convenient Location Adds Luster to Cadence Master-Planned Community," *Las Vegas Sun*, November 10, 2015, www.vegasinc.lasvegassun.com/.

236   "emergency-born community development": "Housing Known as Townsite, Plancor 201-H," U.S. War Assets Administration, 1946, www .hendersonlibraries.sobeklibrary.com/.

237   returned to Lake Mead from the city: "Las Vegas Wash Riparian Restoration Project," Southern Nevada Water Authority, December 9, 2021, www.usbr.gov/.

237   treated in a three-stage process: "Our Wastewater Treatment Process," Clark County Water Reclamation District, September 11, 2015, www .cleanwaterteam.com/.

237   virtually all of it is recycled: "2023 Water Resource Plan," Southern Nevada Water Authority, October 11, 2022, www.snwa.com/.

237   tens of thousands of acre-feet less: Sean Golonka, "Feds Cut Colorado River Water Allocations. What Does it Mean for Nevada?," *Nevada Independent*, August 16, 2022, www.thenevadaindependent.com/.

237   Air Force base in Germany: Abrahm Lustgarten, "The 'Water Witch': Pat Mulroy Preached Conservation While Backing Growth in Las Vegas," ProPublica, June 2, 2015, www.propublica.org/.

238   ordered a temporary moratorium: Interviews with Pat Mulroy on May 17 and August 7, 2023.

238   Nevada would surpass its budget: Robert Reinhold, "Battle Lines Drawn in Sand as Las Vegas Covets Water," *New York Times*, April 23, 1991, www.nytimes.com/.

239   "starting to get squeamish": Michael Weissenstein, "The Water Empress of Vegas," *High County News*, April 9, 2001, www.hcn.org/.

239   one of the wettest decades ever: "Climatic Fluctuations, Drought, and Flow in the Colorado River Basin," U.S. Geological Survey, August 2004, www.pubs.usgs.gov/.

239   fastest-growing city in the nation: "Vegas' Growth Tops in Nation in '90s," *Las Vegas Sun*, December 1, 1999, www.lasvegassun.com/.

239   stockpile in the two reservoirs: "Annual Report," Colorado River Water Users Association, 2002, www.crwua.org/.

240   bottom of the reservoir: Matt Jenkins, "How Low Will Vegas Go for Water?," *High Country News*, June 13, 2005, www.hcn.org/.

241   at least twelve hundred years: Annette McGivney, "Megadrought in the American South-West: A Climate Disaster Unseen in 1,200 Years," *Guardian*, September 12, 2022, www.theguardian.com/.

241   pump groundwater in rural Nevada: Frederic Lasserre, "Water in Las

Vegas: Coping with Scarcity, Financial and Cultural Constraints," *City Territ Archit* 2, no. 11 (2015); www.propublica.org/.

241 "river really goes south": Red Lodge, "Desert Solitaire: Las Vegas Bets Big on Rural Water," *High Country News*, August 15, 2012, www.hcn .org/.

241 lawsuits from a broad array: Ken Ritter, "Judge Again Rejects Vegas Water Pipeline from Rural Valleys," Associated Press, March 10, 2020, www.sltrib.com/.

242 half of the basins that supply freshwater: Sebastien Malo, "US Faces Fresh Water Shortages Due to Climate Change, Research Says," World Economic Forum, March 4, 2019, www.weforum.org/.

242 so much water has been pumped: Mira Rojanasakul, Christopher Flavelle, Blacki Migliozzi, and Eli Murray, "America Is Using Up Its Groundwater Like There's No Tomorrow," *New York Times*, August 28, 2023, www.nytimes.com/.

242 leisure ranchers and the snowbirds: Interviews with Chip Wilson, Bryant Powell, and Al Bravo, August 22, 2022.

243 Superstition Vistas: Grady Gammage et al., "The Treasure of the Superstitions," ASU Morrison Institute for Public Policy, April 2006, www .morrisoninstitute.asu.edu/.

243 annexed the eight square miles: Justin Pazera, "Apache Junction Leader Clarifies Construction on Superstition Vistas," KNXV, April 6, 2022, www.abc15.com/.

243 ideal spot to keep horses there: "Apache Junction Mayor Follows Unique Trail to Leadership," Maricopa Association of Governments, February 18, 2022, www.azmag.gov/.

243 "dipping our first toe in the water": Joshua Bowling, "Developers Are Building in the Massive Superstition Vistas Area Southeast of Phoenix, but Water Will Dictate How It Grows," *Arizona Republic*, February 9, 2022, www.azcentral.com/.

244 consume significantly more water: Jonathan Thompson, "The Colorado River's Alfalfa Problem," *High County News*, September 1, 2022, www .hcn.org/.

244 largest single river user: Janet Wilson and Nat Lash, "The Historic Claims That Put a Few California Farming Families First in Line for Colorado River Water," ProPublica and *Desert Sun*, November 9, 2023, www.propublica.org/.

244 the next seventy-five years: "Colorado River Water Transfer Agreement," San Diego County Water Authority, February 2020, www.sdcwa .org/.

244 municipalities across southern California: "Metropolitan, Palo Verde Boards Approve Colorado River Water Transfer," *WaterWorld*, October 23, 2002, www.waterworld.com/.

244 drip system saved: Mike Hsu, "Drip-Irrigation Study Sees 'Huge' Reduc-

tion in Water, Fertilizer Use for Sweet Corn," University of California Agriculture and Natural Resources, June 16, 2022, www.ucanr.edu/.

244 entirely over to wheat: Abrahm Lustgarten and Naveena Sadasivam, "Holy Crop: How Federal Dollars Are Financing the Water Crisis in the West," ProPublica, May 27, 2015, www.projects.propublica.org/.

245 helping to slow the drain: Shondiin Silversmith, "Gila River Indian Community Receives \$233M in Water Conservation, Infrastructure Funding," *Arizona Mirror*, April 7, 2023, www.azmirror.com/.

245 similar agreements with municipalities: Jake Bittle, "US Turns to Tribes to Help Arizona Survive Colorado River Cuts," *Grist*, April 11, 2023, www.grist.org/.

245 La Paz Valley: Interview with Amelia Flores, March 23, 2022.

247 offered a history lesson: Brenda Burman, "Colorado River 101," presentation at the Colorado River Water Users Association Conference in Las Vegas, December 14, 2022.

247 untouched in times of shortage: John Fleck, "California Wants to Keep (Most of) the Colorado River for Itself," *New York Times*, February 23, 2023, www.nytimes.com/.

249 forgo the \$3.9 billion: Calculated by adding together the total value of agriculture in Imperial, Riverside, and San Bernardino Counties as described in "California Agricultural Statistics Review: 2021–2022," California Department of Food & Agriculture, March 6, 2023, www.cdfa.ca.gov/.

249 One water expert: Interview with Robert Glennon, September 5, 2023.

249 small desalination plant: Rominder Suri et al., "Salton Sea Independent Review Panel Summary Report," Salton Sea Management Program, September 2022.

250 "largest master-planned community": "100,000-Home Teravalis Breaks Ground in Buckeye," Arizona Big Media, December 18, 2022, www.azbigmedia.com/.

250 far West Valley: "Lower Hassayampa Sub-Basin 100-Year Assured Water Supply Projection," Arizona Department of Water Resources, January 2023, www.infoshare.azwater.gov/.

250 eighty-four-thousand acre-feet: Calculated by adding together the 2117 projections for "existing wells," "analysis wells," and "certificate wells" in the "Lower Hassayampa Sub-Basin 100-Year Assured Water Supply Projection," January 2023.

250 entire city of Albuquerque: "Water 2120: Securing Our Water Future," Albuquerque Bernalillo County Water Authority, 2016, www.abcwua.org/.

251 out of compliance with a 1980 law: Joshua Partlow, Yvonne Wingett Sanchez, and Isaac Stanley-Becker, "Phoenix Area Can't Meet Growth Demands Over Next Century," *Washington Post*, June 1, 2023, www.washingtonpost.com/.

251 stalling construction: Glenn Gullickson, "Tartesso Purchased in $80 Million Deal," West Valley View, September 14, 2016, www.westvalleyview.com/.

252 edged it out in the national rankings: Karl Pischke, "The Top-Selling Master-Planned Communities of 2020," RCLCO Real Estate Advisors, January 2021, www.rclco.com/.

AFTERWORD

256 seemingly remote corner: Kay Sutherland, "Rock Paintings at Hueco Tanks State Historic Site," Texas Department of Parks & Wildlife, 2006, www.tpwd.texas.gov/; "Weaving the Story: The People of Hueco Tanks," Texas Beyond History, January 2008, www.texasbeyondhistory.net/.

257 destination for westering migrants: Mabelle Eppard Martin, "California Emigrant Roads Through Texas," *Southwestern Historical Quarterly* 28, no. 4 (1925): 287–301, www.jstor.org/.

257 Silverio Escontrias: "Weaving the Story: The People of Hueco Tanks," Texas Beyond History.

BIBLIOGRAPHY

Abbey, Edward. *Desert Solitaire*. New York: Simon & Schuster, 1968.
———. *The Monkey Wrench Gang*. New York: HarperCollins, 1975.
Alba, Victor. *The Mexicans: The Making of a Nation*. New York: Frederick A. Praeger, Publishers, 1967.
Alcalá, Rosa. *The Lust of Unsentimental Waters*. Bristol: Shearsman Books, 2012.
Alvarez, C. J. *Border Land, Border Water: A History of Construction on the US-Mexico Divide*. Austin: University of Texas Press, 2019.
Anaya, Rudolfo. *Alburquerque*. Albuquerque: University of New Mexico Press, 1992.
Anthes, Bill. *Native Moderns: American Indian Painting, 1940–1960*. Durham: Duke University Press, 2006.
Anzaldúa, Gloria. *Borderlands/La Frontera: The New Mestiza*. San Francisco: Aunt Lute Books, 1987.
Arellano, Juan Estevan. *Enduring Acequias: Wisdom of the Land, Knowledge of the Water*. Albuquerque: University of New Mexico Press, 2014.
Armes, Jay J., with Frederick Nolan. *Jay J. Armes, Investigator: The World's Most Successful Private Eye*. New York: Macmillan, 1976.
Baca, Jimmy Santiago. *Martín and Meditations on the South Valley*. New York: New Directions, 1986.
———. *A Place to Stand*. New York: Grove, 2001.
———. *Singing at the Gates: Selected Poems*. New York: Grove, 2014.
———. *Working in the Dark: Reflections of a Poet of the Barrio*. Santa Fe: Museum of New Mexico Press, 1992.
Banham, Reyner. *Scenes in America Deserta*. Salt Lake City: Gibbs M. Smith, 1982.
Bannerman, Ty. *Images of America: Forgotten Albuquerque*. Charleston: Arcadia Publishing, 2008.

Bégout, Bruce. *Zeropolis: The Experience of Las Vegas.* London: Reaktion Books, 2003.

Berger, Bruce. *A Desert Harvest: New and Selected Essays.* New York: Farrar, Straus and Giroux, 1990.

Binkley, Christina. *Winner Takes All: Steve Wynn, Kirk Kerkorian, Gary Loveman, and the Race to Own Vegas.* New York: Hyperion, 2008.

Bock, Charles. *Beautiful Children.* New York: Random House, 2008.

Bolton, Herbert. *Coronado: Knight of Pueblos and Plains.* New York: McGraw-Hill, 1949.

Bowden, Charles. *Blue Desert.* Tucson: University of Arizona Press, 1986.

———. *Murder City: Ciudad Juárez and the Global Economy's New Killing Fields.* New York: Nation Books, 2010.

Burciaga, José Antonio. *Drink Cultura.* Santa Barbara: Joshua Odell Editions, 1993.

Cabeza de Vaca, Álvar Núñez. *Adventures in the Unknown Interior of America.* Trans. Cyclone Covey. Albuquerque: University of New Mexico Press, 1983.

Canizaro, Vincent, ed. *Architectural Regionalism: Collected Writings on Place, Identity, Modernity, and Tradition.* Princeton: Princeton Architectural Press, 2007.

Cantú, Francisco. *The Line Becomes a River: Dispatches from the Border.* New York: Riverhead Books, 2018.

Cartwright, Gary. *Dirty Dealing: Drug Smuggling on the Mexican Border and the Assassination of a Federal Judge.* El Paso: Cinco Puntos Press, 1998.

Chung, Sue Fawn. *Images of America: The Chinese in Nevada.* Charleston: Acadia Publishing, 2011.

Cockcroft, Eva, John Pitman Weber, and James Cockcroft, eds. *Toward a People's Art: The Contemporary Mural Movement.* Albuquerque: University of New Mexico Press, 1977.

Cornejo Víllavícencío, Karla. *The Undocumented Americans.* New York: One World, 2020.

Cozzens, Gary. *Tres Ritos: A History of Three Rivers, New Mexico.* Charleston: History Press, 2015.

Curtis, William J. R. *Modern Architecture Since 1900.* 3rd rev. ed. New York: Phaidon, 1996.

deBuys, William. *A Great Aridness: Climate Change and the Future of the American Southwest.* Oxford: Oxford University Press, 2011.

DeMark, Judith. *Essays in 20th Century New Mexico History.* Albuquerque: University of New Mexico Press, 1994.

Denton, Sally, and Roger Morris. *The Money and the Power: The Making of Las Vegas and Its Hold on America.* New York: Knopf, 2001.

Dodge, Robert K., and Joseph B. McCullough, eds. *New and Old Voices of Wah'Kon-Tah.* New York: International Publishers, 1974.

Domin, Christopher, and Kathryn McGuire. *Powerhouse: The Life and Work of Judith Chafee*. New Haven: Yale University Press, 2019.

Elsasser, Nan, Kyle MacKenzie, and Yvonne Tixier y Vigil, eds. *Las Mujeres*. New York: Feminist Press, 1981.

Encinias, Miguel. *Two Lives for Oñate*. Albuquerque: University of New Mexico Press, 1997.

Engelbrecht, Lloyd and June-Marie. *Henry Trost: Architect of the Southwest*. El Paso: El Paso Public Library Association, 1981.

Eskeets, Edison, and Jim Kristofic. *Send a Runner: A Navajo Honors the Long Walk*. Albuquerque: University of New Mexico Press, 2021.

Federal Writers' Project, Sponsor. *American Guide Series: New Mexico*. New York: Hastings House, 1940.

Felipe II. *Transcripción de las ordenanzas de descubrimiento, nueva población y pacificación de las Indias*. Ministerio de Vivienda. Madrid: Servicio Central de Publicaciones, 1573.

Fergusson, Erna. *Albuquerque*. Albuquerque: Merle Armitage Editions, 1947.

Flint, Richard. *Great Cruelties Have Been Reported: The 1544 Investigation of the Coronado Expedition*. Albuquerque: University of New Mexico Press, 2002.

Fox, William. *In the Desert of Desire: Las Vegas and the Culture of Spectacle*. Reno: University of Nevada Press, 2007.

Fugate, Francis and Roberta. *Roadside History of New Mexico*. Missoula: Mountain Press Publishing Company, 1989.

George, G. G., and Leigh Conrad. *Images of America: Phoenix's Greater Encanto-Palmcroft Neighborhood*. Charleston: Arcadia, 2014.

Gerdes, Sarah. *Sue Kim: The Authorized Biography*. CreateSpace, 2016.

Gibson, Carrie. *El Norte: The Epic and Forgotten Story of Hispanic North America*. New York: Atlantic Monthly Press, 2019.

Gilb, Dagoberto. *The Last Known Residence of Micky Acuña*. New York: Grove, 1994.

———. *The Magic of Blood*. Albuquerque: University of New Mexico Press, 1993.

Gómez, Laura. *Inventing Latinos: A New Story of American Racism*. New York: New Press, 2020.

Gonzalez, Ray, ed. *After Aztlan: Latino Poets of the Nineties*. Boston: David R. Godine, 1992.

Gonzales-Berry, Erlinda, and David R. Maciel, eds. *The Contested Homeland: A Chicano History of New Mexico*. Albuquerque: University of New Mexico Press, 2000.

Greenough, Sarah, ed. *My Faraway One: Selected Letters of Georgia O'Keeffe and Alfred Stieglitz, Volume 1, 1915–1933*. New Haven: Yale University Press, 2011.

Gregonis, Linda, and Karl J. Reinhard. *Hohokam Indians of the Tucson Basin*. Tucson: University of Arizona Press, 1979.

Gutiérrez, Raquel. *Brown Neon*. Minneapolis: Coffee House Press, 2022.

Hämäläinen, Pekka. *The Comanche Empire*. New Haven: Yale University Press, 2008.

Hansman, Heather. *Down River: Into the Future of Water in the West*. Chicago: University of Chicago Press, 2019.

Hayden, Dolores. *A Field Guide to Sprawl*. New York: W. W. Norton & Company, 2004.

Hernández, Kelly Lytle. *Bad Mexicans: Race, Empire and Revolution in the Borderlands*. New York: W. W. Norton, 2022.

Hershwitzky, Patricia. *Images of America: West Las Vegas*. Charleston: Arcadia Publishing, 2011.

Hickey, Dave. *Air Guitar: Essays on Art & Democracy*. Los Angeles: Art Issues Press, 1997.

———. *Perfect Wave: More Essays on Art and Democracy*. Chicago: University of Chicago Press, 2017.

Hodge, Roger. *Texas Blood: Seven Generations Among the Outlaws, Ranchers, Indians, Missionaries, Soldiers, and Smugglers of the Borderlands*. New York: Knopf, 2017.

Hoig, Stan. *Came Men on Horses: The Conquistador Expeditions of Francisco Vázquez de Coronado and Don Juan de Oñate*. Boulder: University Press of Colorado, 2014.

Jackson, John Brinckerhoff. *A Sense of Place, a Sense of Time*. New Haven: Yale University Press, 1994.

Johnson, G. Wesley Jr., ed. *Phoenix in the Twentieth Century: Essays in Community History*. Norman: University of Oklahoma Press, 1993.

Kang, Jay Caspian. *The Loneliest Americans*. New York: Crown, 2021.

Kessell, John. *Kiva, Cross, and Crown: The Pecos Indians and New Mexico 1540–1840*. Washington, D.C.: National Park Service, 1979.

Koch, Natalie. *Arid Empire: The Entangled Fates of Arizona and Arabia*. London: Verso, 2022.

Kuhn, Eric, and John Fleck. *100 Years Ago in the Negotiations of the Colorado River Compact*. Albuquerque: University of New Mexico School of Law Utton Transboundary Research Center, 2022.

Layne, Ken. *Desert Oracle Volume 1: Strange True Stories from the American Southwest*. New York: Farrar, Straus and Giroux, 2020.

Littlejohn, David, ed. *The Real Las Vegas: Life Beyond the Strip*. Oxford: Oxford University Press, 1999.

Locke, Raymond Friday. *Book of the Navajo*. Los Angeles: Mankind Publishing, 1976.

Logan, Michael. *Desert Cities: The Environmental History of Phoenix and Tucson*. Pittsburg: University of Pittsburg Press, 2006.

Long Soldier, Layli. *Whereas*. Minneapolis: Graywolf, 2017.

López-Stafford, Gloria. *A Place in El Paso: A Mexican-American Childhood*. Albuquerque: University of New Mexico Press, 1996.

Lovato, Roberto. *Unforgetting: A Memoir of Family, Migration, Gangs, and Revolution in the Americas.* New York: Harper, 2020.

Luckingham, Bradford. *Minorities in Phoenix: A Profile of Mexican American, Chinese American, and African American Communities, 1860–1992.* Tucson: University of Arizona Press, 1994.

———. *The Urban Southwest: A Profile History of Albuquerque—El Paso—Phoenix—Tucson.* El Paso: Texas Western Press, 1982.

Luiselli, Valeria. *Lost Children Archive.* New York: Knopf, 2019.

Lummis, Charles. *The Land of Poco Tiempo.* New York: Charles Scribner's Sons, 1893.

Marshall, James. *Santa Fe: The Railroad That Built an Empire.* New York: Random House, 1945.

Martínez, Elizabeth Sutherland, and Enriqueta Longeaux y Vásquez. *Viva la Raza!: The Struggle for the Mexican-American People.* New York: Doubleday, 1974.

Martínez, Óscar. *The Beast: Riding the Rails and Dodging Narcos on the Migrant Trail.* London: Verso, 2013.

Martínez, Rubén. *Desert America: Boom and Bust in the New Old West.* New York: Metropolitan Books, 2012.

McLuhan, T. C. *Dream Tracks: The Railroad and the American Indian 1890–1930.* New York: Abrams, 1985.

McMaster, Gerald, ed. *Reservation X: The Power of Place in Aboriginal Contemporary Art.* Seattle: University of Washington Press, 1998.

McPhee, John. *Encounters with the Archdruid.* New York: Farrar, Straus and Giroux, 1971.

Mora, Pat. *Chants.* Houston: Arte Público Press, 1984.

Needham, Andrew. *Power Lines: Phoenix and the Making of the Modern Southwest.* Princeton: Princeton University Press, 2014.

Neff, Emily Ballew. *The Modern West: American Landscapes 1890–1950.* New Haven: Yale University Press, 2006.

Orleck, Annelise. *Storming Caesars Palace: How Black Mothers Fought Their Own War on Poverty.* Boston: Beacon Press, 2005.

Otero, Lydia. *In the Shadows of the Freeway.* Tucson: Planet Earth Press, 2019.

———. *La Calle.* Tucson: University of Arizona Press, 2010.

Owen, David. *Where the Water Goes.* New York: Riverhead, 2017.

Paskus, Laura. *At the Precipice: New Mexico's Changing Climate.* Albuquerque: University of New Mexico Press, 2020.

Pearce, T. M. *New Mexico Place Names: A Geographical Dictionary.* Albuquerque: University of New Mexico Press, 1965.

Powell, Dana. *Landscapes of Power: Politics of Energy in the Navajo Nation.* Durham: Duke University Press, 2018.

Prat, Ramon, Michael Kubo, Irene Hwang, and Jaime Salazar, eds. *Desert America: Territory of Paradox.* Barcelona: Actar, 2006.

Reid, Ed, and Ovid Demaris. *The Green Felt Jungle*. New York: Trident Press, 1963.

Reisner, Marc. *Cadillac Desert: The American West and Its Disappearing Water*. New York: Penguin, 1986.

Reséndez, Andrés. *The Other Slavery: The Uncovered Story of Indian Enslavement in America*. New York: Houghton Mifflin Harcourt, 2016.

Roberts, David. *The Pueblo Revolt: The Secret Rebellion That Drove the Spaniards Out of the Southwest*. New York: Simon & Schuster, 2006.

Ross, Andrew. *Bird on Fire: Lessons from the World's Least Sustainable City*. Oxford: Oxford University Press, 2011.

Rothstein, Richard. *Color of Law: A Forgotten History of How Our Government Segregated America*. New York: W. W. Norton & Company, 2017.

Sando, Joe, ed. *Po'Pay: Leader of the First American Revolution*. Santa Fe: Clear Light Publishing, 2005.

Schipper, Janine. *Disappearing Desert: The Growth of Phoenix and the Culture of Sprawl*. Norman: University of Oklahoma Press, 2008.

Scott, Nancy. *Georgia O'Keeffe*. London: Reaktion Books, 2015.

Scott Brown, Denise, Robert Venturi, and Steven Izenour. *Learning from Las Vegas*. Cambridge: MIT Press, 1972.

Selnow, Gary, and Sam G. Riley. *Regional Interest Magazines of the United States*. Westport, CT: Greenwood Publishing Group, 1991.

Sheridan, Thomas. *Arizona: A History*. Rev. ed. Tucson: University of Arizona Press, 2012.

Shreve, Bradley. *Red Power Rising: The National Indian Youth Council and the Origins of Native Activism*. Norman: University of Oklahoma Press, 2011.

Silko, Leslie Marmon. *Ceremony*. New York: Penguin, 1977.

Simmons, Marc. *Albuquerque: A Narrative History*. Albuquerque: University of New Mexico Press, 1982.

———. *The Last Conquistador: Juan de Oñate and the Settling of the Far Southwest*. Norman: University of Oklahoma Press, 1991.

Sklair, Leslie. *Assembling for Development: The Maquila Industry in Mexico and the United States*. New York: Routledge, 1989.

Smythe, William Edward. *The Conquest of Arid America*. New York: Harper & Brothers, 1900.

Tejada, Roberto. *Still Nowhere in the Empty Vastness: History + Metaphor*. Blacksburg, VA: Noemi Press, 2019.

Thompson, Hunter. *Fear and Loathing in Las Vegas: A Savage Journey to the Heart of the American Dream*. New York: Random House, 1972.

Timmons, W. H. *El Paso: A Borderlands History*. El Paso: Texas Western Press, 1990.

Treib, Marc. *Sanctuaries of Spanish New Mexico*. Berkeley: University of California Press, 1993.

Trimble, Marshall. *Roadside History of Arizona*. Missoula: Mountain Press Publishing Company, 1986.

Tronnes, Mike, ed. *Literary Las Vegas.* New York: Henry Holt & Company, 1995.

Vint, Bob, and Christina Neumann. *Southwest Housing Traditions: Design, Materials, Performance.* Washington, D.C.: U.S. Department of Housing and Urban Development Office of Policy Development and Research, 2005.

Wheeler, Joshua. *Acid West: Essays.* New York: Farrar, Straus and Giroux, 2018.

Whitaker, Matthew. *Race Work: The Rise of Civil Rights in the Urban West.* Lincoln: University of Nebraska Press, 2007.

Wright, Frank Lloyd. *In the Realm of Ideas.* Carbondale: Southern Illinois University Press, 1988.

Zepeda, Ophelia. *Where Clouds Are Formed.* Tucson: University of Arizona Press, 2008.

INDEX

Abbey, Edward, 69
Abernathy, Ralph, 139
Abilene, 95
Abiquiú village, 26
Acoma Pueblo, 16–17, 226–27, 258
Acequia Madre, 208
Adams, Ansel, 46, 60
adobe, 51, 86–88, 173
Afghanistan War, 210, 218
AFL-CIO, 142
African migrants, 211
agriculture, xxii–xxiii, 47–52, 55, 57, 138, 176, 223–24, 244–45, 249, 256
Ajo, Arizona, 204
Akimel O'odham, 49–51, 230, 245
Alabama, 143
Aladdin casino, xx
Alameda Land Grant, 157–58
Alamogordo, 199–200
Alaska, 33, 130
Albuquerque, xiv, xvi–xvii, xx–xxi, 3–7, 10–13, 24–31, 35–40, 43–45, 66, 111, 123, 137, 157–77, 191, 208, 218–19, 224–31, 239, 246, 250, 252
  See also Villa de Alburquerque
Albuquerque Academy, 5
Albuquerque High School, 38
Albuquerque Indian School, 160–61
Albuquerque Journal, 161, 166, 170, 227–28

Albuquerque Police Department, 168–70, 174
Albuquerque Tribune, 168–69
Alcalá, Rosa, 201
Alcalde, New Mexico, 225–26, 228–29
Alcanfor Pueblo, 11, 13
Ali, Muhammad, 133
Alianza Federal de Mercedes, 165
Alianza Hispano-America, 75
Alibi, 4, 225–26
Allen, George, 36
Allen, William, 36
Altar Desert, 202
Alvarado, Hernando de, 10–11, 33
American Book Award, 176
American Canal, 208
American Smelting and Refining Company (ASARCO), 100, 112, 116
AMREP, 158
Angelou, Maya, 176
Anglos, xv–xvi, xxiii, 5, 24, 26–27, 32–33, 35–37, 39, 48–50, 66, 74–75, 95, 99–100, 158, 163–65, 170, 172, 190, 219, 245
Animas River, 234
Annunciation House, 213–15
Anthony, Saint, 114

Antonio J. Bermúdez Industrial Park,
    103
Anza-Borrego, xiii
Anzaldúa, Gloria Desert State Park,
    4, 163, 219
AP, 107, 169–70
Apache, 5, 11, 17–18, 21, 23, 25, 27,
    36, 67, 73, 86, 88, 163, 257
Apache Junction, 44, 67, 242–44, 246,
    251
architecture, 79–91, 127, 188–92,
    199
Arenal massacre, 12–13
Argüello Carvajál, Fernando de, 18
Arizona, xvii, xxi–xxii, 76, 58, 224.
    See also Phoenix; Tucson; and other
    specific locations
    agriculture and, xxiii, 47–51
    Anglos and, 48–51, 219
    Apache and, 73
    archaeology and, 70
    architecture and, 79–85, 188–91
    Blacks and, 184–85
    border and migrants and, 5, 200–207
    electricity and, 42, 188
    growth and, xviii, 45–47, 59–65, 127,
        250
    Treaty of Guadalupe Hidalgo and,
        26, 164
    water and, 51–55, 242–50
    welfare and, 140–41
Arizona Canal, 64–65
Arizona Daily Star, 79, 207
Arizona Department of Water
    Resources, 195, 250
Arizona Highways, 43–46, 58–65,
    126–27
Arizona Highways Department, 58
Arizona Historical Society, 65
Arizona Republic, 56, 82, 186, 193
Arizona State University (ASU), 62,
    186, 243
    Morrison Institute, 243
Armes, Jay J. (Julian Armas), xxii,
    105–11
Artspace, 79

Ashland, Oregon, 222
Asian Americans, 184–85, 219–21,
    224
Asian Community Development
    Council of Nevada, 221
Asian immigrants, 219, 231
Atchison, Topeka and Santa Fe
    Railway (AT&SF), 29–31, 35–36,
    43, 169
Atlanta, 77
Atrisco, New Mexico, 24
Austin, George, 87
Austin, Texas, 95, 113
Australian Aborigines, 71
Avey, George, 64
Avila, María, 137
Aztecs, 70, 256

Baca, Jimmy Santiago, 159–62, 167,
    170–77
Baca, Mieyo, 160, 162
Backel, Ethel, 38
Bailey, Anna, 134, 136
Baja California, 53
Ball, Tom, 122, 131
Banham, Reyner, 97
Bank of Las Vegas, 147
Barnes, Edward Larrabee, 79
Barrio Bread bakery, 221–24
Bartlett Dam, 58
Baruch College, xviii
Bégout, Bruce, 123, 155
Belafonte, Harry, 135–36
Belen, New Mexico, 26
Bellagio, 148–49, 153, 237
Bernalillo County, New Mexico,
    10–11, 24, 29
Biden, Joe, 212, 214–16, 218, 251
Bigotes, 9–11, 13
Biklen, John, 83–84, 90
Bimson, Walter, 59, 63, 126–28, 147,
    183, 185
Binion, Benny, 128, 144, 148
Binion's Horseshoe, 144
Binkley, Christina, 149
Bishop's Lodge, 53–54

Black Berets, 168, 170
Black Lives Matter, 73
Black Panthers, 168
Blacks, xvii, 5, 75, 77, 133–36, 138, 143, 163, 166, 168, 182–86
Black Wall Street, 182
Blood In, Blood Out (film), 176
Border Patrol, 114, 199–200, 203, 205, 207, 211, 213–15. *See also* Mexican border wall; migrants
Bosque Redondo, Long Walk to, 27, 41
Boston, xx–xxi, 77, 149, 227, 231
Boulder Canyon, 258
Boulder City, 125, 238
Boulders Resort, 222
Bowden, Charles, 159
Bracero Program, 103
Bradford, James, 168
Brando, Christian, 106–8
Brando, Marlon, 106–8
Brazilian migrants, 214
Brown, Denise Scott, xvii, 122
Brown Derby, 135
Bryan, R. W. D., 161
Buckeye, Arizona, 44, 250–52
Buffalo Soldiers, 258
Burciaga, José Antonio, 101
Bureau of Reclamation, 52–55, 125, 247
Burman, Brenda, 247
Bush, George W., 218
Butterfield Overland Mail, 257

Cabeza de Vaca, Álvar Núñez, xv–xvi, 7
Cadence subdivision, 235–36, 252
Caesars Palace, xix, xx, 139, 149, 153, 155, 246, 248, 250
Cahuilla people, xiv
Caldwell, R. E., 54
Calhoun, John C., 31
California, xviii, 14, 26, 28–29, 53–55, 83, 102, 106, 137, 141, 150, 172, 194, 201, 244–49
California Volunteers, 48

Camelback Mountain, xv, 43, 188
Camino Real de Tierra Adentro, 99
Canales, Rito, 170
Cantú, Francisco, 207
Canty, Hattie, 134, 143
Canyon de Chelly, 27, 41
Capone, Al, 126, 129
Carano, Anthony, 155
Cárdenas, García López de, 11–13
Cardis, Louis, 96
Carleton, James, 26–27
Carlisle Indian School, 161
Carlsbad, New Mexico, 190
Carlson, Raymond, 46, 57–64
Carpenter, Delph, 53–54
Carson, Kit, 36
Carson City, Nevada, 124, 237
Cartwright, Gary, 108–10
Casa Caldera, 191
Casa del Migrante, 211
Casa Grande, Arizona, 69–70
Casteel, Bette, 39
Cataract Canyon, 53
Catawba tribe, 165
Caterpillar corporation, 78
Catholic Church, 101, 112–13, 206, 213
Catron, Thomas B., 164
Census, 163, 172, 231
Centers for Disease Control, 116
Central America, xxii–xxiii, 69, 173
Central American migrants, 201–6, 211, 216–18
Central Arizona Project, 186, 241, 244, 247
Central Michigan University, xviii
Cerro de los Muleros (*later* Mount Cristo Rey), 99, 101, 112
Chaco Canyon, xv, 7–8
Chafee, Judith, 79–93, 191–92, 252
Chagra, Jimmy, xix–xx
Chagra, Lee, xix–xx
Chama, New Mexico, 164
Chaney, Willa Mae, 132
Chapel in the Valley mortuary, 183
Chavez, Cesar, 139

Chávez, Marty, 5, 226
Chávez, Ventura, 163
Chemehuevi tribe, 245
Cherokee tribe, 163, 165
Chesapeake Bay, 242
Chicago, 45, 59, 63, 75, 79, 92, 126, 129, 130, 214, 231
Chicago Tribune, 32, 107
Chicano activists 164–76. *See also* Mexican Americans
uprising of 1971, 168–70, 174
Chihuahua, 14, 28, 100, 112
Chihuahuan Desert, xiii–xiv, 7, 95, 115, 199, 200, 256
Chinese immigrants, xxiii, 75, 87, 220, 231
Cibola National Forest, 4
Cicuye (*later* Pecos Pueblo), 9–10
Circa casino, 151–56
Circus Circus, 139
Cisneros, Sandra, 176
Civil War, 27, 32, 48, 164
Clark, Donald Maurice, 134
Clark, Wilbur, 129
Clark, William, 124–25
Clark County, Nevada, 130, 142, 220, 237–42
Clark County Welfare Rights Organization, 138–40
Cli, Peshi, 37
cliff dwellings, xv, 7–8, 256
climate change, xxi–xxii, 44, 188–89, 197–98, 217, 220, 259
Clinco, Demion, 85–86
Clinton, Bill, 5, 140–41, 206
Clinton, Hillary, 143
Clooney, Rosemary, 135
Coalition for Navajo Liberation, 165
Cochiti Pueblo, 19, 23
COESPO (Chihuahua immigration agency), 210–12, 216
Colombia, xix
Colorado, 15, 26, 164, 233–34
Colorado City, Texas, 95
Colorado Plateau, xiv, 7–8, 26–27, 48, 165, 190, 256

Colorado River, 53–56, 124–25, 186, 190, 234, 237–41, 244–49
Colorado River Compact, 53–55, 246–47
Colorado River Indian Tribe, 245
Colorado River Water Users Association, 246–50
Colston, Edward, 227
Columbus, Christopher, 163, 167, 227
Comanche, 5, 25–26, 73, 167, 190
Comstock Lode, 124
Confederate Army, 26–27, 49
Conixu, Luis, 19
Conquest of Arid America, The (Smythe), xvi, 33, 50–51, 218
Constantine, Emperor, 113
Continental Bank, 147
Contras, 206
copper mining, 57, 124–25
Corbett, Jim, 205–6
Corbusier, 81–82, 85
Cordova, Antonio, 170
corn, xxii, 47–48
Cornero, Tony, 129
Coronado, Francisco Vázquez de, 9–17, 22–23, 35–36
Coronado Historic Site, 13
Cortés, Hernán, 14, 33
Cosby Show, The (TV show), 154
Cosgriff, Walter, 127, 147
Costa, Lourdes, 112–13, 115–16
cotton, xxii–xxiii, 69–70, 134, 195, 245
COVID-19 pandemic, 196, 212
Cowabunga Bay Water Park, 235, 237
Crear, Cedric, 141
Crosby, Bing, 63
Crypto Jews, 163
Cuarto Centenario (Albuquerque), 225–26
Cuba, 33
Cuban migrants, 211–12, 214
Cuervo y Valdés, Francisco, 24
Culiacán, Mexico, xvi, 7, 9
Culinary Workers Union Local 226, 142–46

Curtis, William J. R., 81–82
Customs and Border Protection, 215

Dae Han sisters, 132
Dalitz, Moe, 147
Dallas, 60, 95
Davis, Ivor, 107
Davis, Sammy, Jr., 135, 140
Davis-Monthan Air Force Base, 89
D casino, 150, 154
DeConcini, Dennis, 206
de la Cruz Jonva, Nicolas, 19
Democratic Party, 58, 228
Denver, 39
Denver Nuggets, 152
Depression, 39, 57, 59, 112, 127
desalination plants, 242, 249–50,
    252
Deseret National Bank, 147
Desert Inn, 129, 147, 149
Desert Rose Motel, 237
Desert Sands Motel, 162
Detroit Free Press, 107
Detroit Lions, 155
Devoid, Alex, 207
Díaz, Porfirio, 100
Diné language, 37–38, 71
Diné people. *See* Navajo
Dinétah, 26–27
Domin, Christopher, 84–85
Domínguez, Francisco Atanasio, 25
Dominy, Floyd, 43
Don Quixote (Cervantes), 28
Dragon's Back, 157
Drake, Sir Francis, 14
drought, xxi, 8, 18, 39, 52, 55–56, 73,
    194, 241–42, 244, 246–47. *See also*
    water
Drought Contingency Plan, 247
Dry Lake project, 42
Dubai climate conference, 230
Duke University Blue Devils, 154
Duncan, Roy, 136
Duncan, Ruby, 134–39
Dunes Hotel and Casino, 129, 147
Duppa, Darrell, 49

Durante, Jimmy, 128
Dust architects, 190–91

Earl, Robert, 91
Earp, Wyatt, 36
eastern European migrants, 211
East Mountains, New Mexico, 5
Edelson, Norm, 92
Ed Sullivan Show, The (TV show), 130
Edward II (Marlowe), 195
Egypt, ancient, 47
Einstein, Albert, 36
Ellin, Nan, 195
Ellis Island, 231
El Morocco, 135
El Movimiento, 167
El Paso del Norte, xvi–xvii, xix–xxi,
    21–22, 95–117, 200, 206, 208–14,
    218–19, 239, 255, 257–58
El Paso Herald-Post, 110
El Paso Matters, 102, 214
El Paso Times, 102–3, 105–6, 110
El Salvador, 204, 206, 212, 216
Emerson, Ralph Waldo, 177
Encanto Village, Phoenix, 64
Entrada pageant, 229
Entsminger, John, 240–41, 248
Environmental Protection Agency
    (EPA), 186
Erdoğan, Recep Tayyip, 217
Escontrias, Silverio, 257
Española, New Mexico, 229–30
ESPN, 154
Estancia, New Mexico, 159–61, 171
Estevánico, 7–9
Estrella Mountains, 44, 48
Export Processing Zone, 103

Fair Housing Act (1968), 186
Farah Manufacturing Company,
    103–4
FBI, 133
Fergusson, Erna, 37
Férnandez de la Cueva, Francisco, 24
Fiesta del Sol, 62
Fiesta de Santa Fe, 22–23

Fife, John, 204–7
financial crisis of 2008, 158, 175, 193, 251, 256
First Christian Church (Tucson), 207
Fitzgerald casino. *See* D casino
Flagstaff, Arizona, 43, 222–23
Flamingo Hotel, 126–29, 136, 139, 147, 183
Flinn, Monica, 81
Florence Correctional Center, 68, 172
Flores, Amelia, 245
Florida, xv, xviii, 126, 150
Floyd, George, 227
Ford, John, 60
Fort Bliss, 103, 111
Fort Defiance, 27
Fort McDowell, 48
Fort Sumner, 27
Fortune, 147
Fountain Hills, 44
Four Queens casino, 150
Francis, Saint, 20
Franciscans, 9, 15, 17–21, 228
Franklin Mountains, xv, 98–99, 200
Freedom Seder, 205
Frenchman Mountain, xv
Frey, Albert, 83
Friends (TV show), 154
Frolander, Danielle, 141
Frontier Hotel and Casino, 143–44, 146

Gabriel, Juan, 101
Gadsden Purchase, 50, 74
Galisteo Basin pueblos, 16, 19
Gallego, Kate, 192
Galveston, xvi
Garcia, Bobby, 168
Garcia, Ruben, 213–15
Gay and Lesbian Latinos Unidos, 77
Genízaros, 18, 25–26
Gerdes, Sarah, 131
German immigrants, 28–29
Geronimo, 86
Getty Center, 93

GI Bill, 185
Gila River, 47–48, 50, 68–69, 74
Gila River Indian Community of Akimel O'otham and Pee Posh, 245
Gilb, Dagoberto, 111–12
Gilgamesh, Epic of, 47
Glassman, Dora, 38
Glassman, Ethel Backel, 38
Glassman, Nathan, 38
Glassman, Sandra, 38–39
Glen Canyon Dam, 190, 234, 246, 249
Glenn, Casper, 205
Glitter Gulch casino, 153
Golden Gate casino, 153–54
Golden Nugget casino, 147–49
Goldfield, Nevada, 125
gold mining, 28, 48–49, 67–68, 245
Goldwater, Barry, xxii, 46, 59, 62–63, 183
Goldwater, Robert, 59, 62–63, 183
Gonzales, Javier, 22
Gonzales, Juan, 158
Good Night America (TV show), 105–6, 108
Gordon, Phil, 192
Grand Canyon, *xi*, xxi, 53–54, 222
Great Basin, xiv
Great Plains, 25, 200, 242
Great Plains Indians, 10
Great Recession, 102, 243
Greek immigrants, 183
Greenbaum, Gus, 128–29
Green Bay Packers, 155
Grey, Zane, 60
Gropius, Walter, 79
Guadalajara, 14
Guadalupe Hidalgo, Treaty of (1848), 25–26, 28, 32, 95, 99, 116, 163, 210
115th anniversary, 164
Guadalupe Mountains, xiii, 95
Guaidó, Juan, 214
Guatemalan migrants, 216–17
Guerra, Don, 221–24
Guerra, Salvador de, 18
Guerrero, John, 103–4

Guevara, Che, 73
Guggenheim, Harry, 36, 100, 116
Gutiérrez, Raquel, 87

Haaland, Deb, xxii, 228–29
Haitian migrants, 217
Hale, Kenneth, 71
Harrah's casino, 132
Harvey, Fred, Company, 35–36
Ha:ṣañ Preparatory & Leadership
    School, 72
Hashknife Pony Express, 65
Havasu Canyon, 60
Hawaii Five-O (TV show), 110
Hawikuh settlement, 7–9, 11
Hayden, Carl, 63
Hayes, Cade, 190–92
Haywoode, Terry, xviii–xix
Head, Robert, 105
Health, Education, and Welfare
    Department, 139–40
Heard, Dwight, 59–60, 63, 183–84
Heard Museum, 184
Hearst, William Randolph, 37–38,
    128
Hefner, Hugh, 130
Helton, Sandra, 92
Henderson, Esther, 60–61
Henderson, Nevada, 235–38, 252
Hermosillo, 73, 88
Hernández, Francisca James, 111
Hernandez, Margo, 79
Herrscher, H. Harry, 183
Hertz, Edwin, xviii, xix
Hickey, Dave, 145, 156
Hickok, Wild Bill, 36
High County News, 239
Highland Elementary School (Las
    Vegas), 133
Hill, Jonah, 90
Hilton Hotel (El Paso), 101
"Hispanic," as term, 175–76. *See
    also* Central American migrants;
    Chicano activists; Mexican
    Americans; *and specific national
    groups*

Hispano gentry, 25, 29–30, 39
Historic Westside Legacy Park (Las
    Vegas), 133–34, 140
Hobbs, Katie, 251
Hohokam culture, xv, 47–48
Hohokam villages, 48
Holbrook, Arizona, 41, 65
Holden, Win, 45
Holmes, Brownie, 68
Holmes, Dick, 68
Home Owners' Loan Corporation
    (HOLC), 182
Honduran migrants, 217
Honduras, 212, 216
Honeywell, 63, 104
Hoodenpyl, George, 54
Hoover, Herbert, 54–55, 125, 246
Hoover, Robin, 207
Hoover Dam, 125, 190, 237, 246,
    249, 258
Hope, Bob, 63, 135
Hopi, 8, 11, 17–18, 20, 36, 69, 190,
    230, 245
Hopi language, 71
Hornbuckle, Bill, 155
Horton, D. R., 251
Hosteen (Navajo man), 37
House on Mango Street, The (Cisneros),
    176
Houston, 89
Howard, Charles, 95, 96
Howard Hughes Corporation, 250
Hualapai tribe, 54
Hueco Tanks State Park, 255–59
Huezo, Rosa, 201, 204
Hughes, Howard, 146–47
Humane Borders, 206–7
Huning, Franz, 28–31, 37, 252
Huning Castle, 31, 252

Iberian heritage, 163–64
IGT, 153
Immigration and Customs
    Enforcement (ICE), 68
Immigration and Naturalization
    Service (INS), 205

Imperial Irrigation District (IID), 244, 248
Imperial Valley, California, 200
Indian activists, 165, 172, 229–31
Indian boarding schools, 160–61, 228
Indian Citizenship Act (1924), 32
"Indian Detour" train, 35–36
Indian House, 184
Indians (Indigenous people; Native Americans), xv–xvii, 5, 8, 10–11, 17–18, 35–38, 49–55, 62, 69–70, 72, 82, 86, 88–89, 158, 160–66, 182, 184–85, 190–91, 204, 224–31, 245. See also Puebloans; and specific peoples; and tribes
Indian voting rights, 32
Indigenous languages, 71–72
Indigenous Wisdom Keepers, 230
Indo-Hispanic descent, 163–64
Interior Department, 228, 247
International Boundary Commission, 218
International Hotel, 147
International Organization for Migration, 212
Investigators PI shop, 106–7, 110
Iraq War, 210, 218
irrigation canals (acequia), xxii, 6, 39, 43, 48–52
Isleta Pueblo, 10, 23, 32, 40, 99, 105
Italian immigrants, 77, 231

Jackson, Jesse, 144
Jackson, Sharon, 90
Jackson, Stonewall, 49
Jackson, Woody, 90
Jackson-Reed Act (1924), 231
Jacobson, Joan, 85
Jacobson House, 84–85
Jake the Barber, 129
Japanese American Citizens League, 220
Jemez Mountains, 16, 23, 29
Jemez Pueblo (Walatowa), 18–20
Jemez River, 13

Jewish immigrants, 28–29, 77, 231
Jiles, Trina, 132
Jim Crow, 163, 182
Johns, Jacob, 230
Johnson, Lyndon, 103
Johnson, Philip, 79
Johnson & Johnson, 111
Joint Chihuahua Operation, 210
Jornada Mogollon people, 256–57
Joseph, Saint, 114
Juan de Jesús, 20–21
Juárez, Benito, 100
Juárez, Mexico, 97–101, 103–5, 111, 115–16, 200, 208–13, 216–19
Juárez Cartel, 210
Juárez Mountains, 98
Jumano people, xiv
Justice Department, xx

Kahn, Louis, 79
Ka-'p-geh Pueblo, 19
Kashfi, Anna, 106–7
Katsina rituals, 8, 18, 20, 23, 256
Kearney, Stephen, 25, 30
Kelly, Mark, 249–50
Kent State University, 168
Keres people, 11, 19
Kerkorian, Kirk, 147, 149, 155
Kesha, 152
Kightlinger, Jeffrey, 248
Kim, Ai-Ja, 121–22, 130–32, 150
Kim, Jane, 132
Kim, Min-Ja "Mia," 121–22, 130–32
Kim, Sook-Ja "Sue," xxii, 121–22, 130–32, 146, 150, 220
Kim Hae-Song, 121–22
Kirtland Air Force Base, 39–40, 168
Kiva Club, 165
Koontz Ranch, 158
Korean Americans, 220
Korean Kittens, 131
Korean War, 122
Kuapoga Pueblo (later Santa Fe), 17
Kuaua Pueblo (later Coronado Historic Site), 13

La Bestia (migrant train), 216–17
labor unions, 142–45
La Calle (Otero), 78
Laguna Mountains, xiii
Laguna Pueblo, 228
Laidlaw, Don, 76
La Jornada (Sabo and Rivera), 224–29
Lake Mead, 190, 236–41, 245–47, 249, 258
Lake Powell, 234, 247
Lake Roosevelt, 56
Lake Tahoe, 130–31
Lamarque, Libertad, 75
La Mujer Obrera, 111
Lamy, Jean-Baptiste, 54
Land of Poco Tiempo. The (Lummis), 34
Lansky, Meyer, 126, 147
La Paz Valley, 245
La Plaza Theatre, 75
Larkin, Lloyd, 184
Las Cruces, New Mexico, 109, 199
Last Known Residence of Mickey Acuña, The (Gilb), 111–12
Las Trampas, New Mexico, 26
Las Vegas, xvi–xvii, xix–xxi, 36, 100, 121–56, 219–21, 236–42, 246–48, 251
Las Vegas Evening Review-Journal, 125
Las Vegas Hilton, 132, 149
Las Vegas Metropolitan Police Department, 133
Las Vegas Raiders, 221
Las Vegas Strip, 121–22, 126–27, 129–30, 134–41, 146, 148–49, 153, 155
 integration of, 134–36, 138
 labor unions and, 144–45
 welfare rights and, 138–40
Las Vegas Sun, 121
Las Vegas Valley, 123–25, 235
Las Vegas Wash, 236–37, 239
La Voz Del Pueblo, 33
Lazarus, Emma, 230–31
Leadville, Colorado, 57
Learning from Las Vegas (Brown and Venturi), xvii

Lebanese immigrants, xix
Lee company, 111
Lee's Ferry, 54
Lerma, Antonio, 158
Levi's company, 111
Life, 71, 135
Lil Jon, 150
Lin, Vida, 220–21
Little Colorado River, 27
Little San Bernardino Mountains, xiii
Llano Estacado, 159
Lockheed Martin, 63, 108
London, Jane, 82, 90
Long, John F., 185–86
Long, Mary, 185
Long Beach, California, 54
Long Soldier, Layli, 3
Los Angeles, 38, 59, 75, 77, 89, 103, 106, 128, 131, 133, 149, 201–3, 208, 243, 246, 248
Los Angeles Lakers, 152
Los Angeles Police Department (LAPD), 125–26
Los Angeles Times, 84
Lost Dutchman Mine, 67–68
Louis, Joe, 135
Lovato, Roberto, 204
Lubbock, Texas, 190
Lucero, Dolores, 163
Luckingham, Bradford, 51
Luke Air Force Base, 182
Lummis, Charles Fletcher, 33–34, 36
Luna, Patricia, 166

Maduro, Nicolás, 214
MAGA movement, 229–30
Malcolm X, 132
Manifest Destiny, 52
Manso people, xv, 99
Manuelito (Navajo leader), 27
maquiladoras, 103, 105, 111, 209
Marble Canyon, 53
Marble Mesa (Henderson), 235
Maricopa County, Arizona, 196–98
Marley, Jennifer, 22–23, 229
Marlowe, Christopher, 195

Martín and Meditations on the South Valley (Baca), 175
"Martín" (Baca), 170–72
Martínez, Óscar, 216–17
Martínez, Rubén, 181
Maxim Hotel, 143
Mazatzal Mountains, 42, 52
McAfee, Guy, 125
McClure, W. F., 54
McCurdy, Chase R.
  Living Black Pillars, 133–34
McDowell Mountains, 56, 65, 188
McGuire, Kathryn, 82–85, 90–93
McGuire Sisters, 121
McMillan, James, 136, 138
Meem, John Gaw, 184, 191–92
Meier, Richard, 93
Mentuhotep II, 47
Mesa, Arizona, 193, 242
Mesa Verde, xv, 7
Mescalero Apache, 27, 257
Methodist Church, 212
Metropolitan Water District of
  Southern California, 246
Metro Squad, 168, 170
Mexicali, 103
Mexican Americans, xvii, xxii–xxiv,
  5, 9–10, 17–18, 24, 32, 34, 40, 51,
  74–78, 86–89, 95, 103–4, 160–76,
  182–87, 207, 217, 219, 224. *See also*
  Chicano activists
Mexican-American War (1846–48), 5,
  25–28, 30, 74, 158, 164, 257
Mexican Army, 209–10
Mexican border wall, 98, 114–16, 218.
  See also migrants
Mexico, xiv, 31, 33, 74, 96, 100,
  102–3, 165, 173, 217. *See also*
  Juárez; *and other specific locations*
  independence from Spain, 25, 73
  Revolution, 99
  Treaty of Guadalupe Hidalgo and,
    25–26, 28, 33
  water rights and, 53, 55
Mexico City, xv–xvi, 17, 99, 201
MGM Grand, 149, 155

Miami, Arizona, 57
Miami Herald, 107
Miami Silver Belt, 57
Michigan State University Spartans,
  154
Michler, Nathaniel, 50–51
Migrant Protection Protocols (MPP),
  212, 216
migrants, 5, 68, 191, 201–19
Miller, George, 137–40
Ming and Ling, 122
Mirage Hotel, 148–49, 153
Mississippi, 134, 182
Mississippi River, 250
Missouri, 65
Missouri River, 250
Moakley, Joe, 206
Modernism Week, 85
Mogollon people, 256
Mogollon Rim, 41, 48
Mohave tribe, 245
Moho Pueblo, 13
Moïse, Jovenel, 217
Mojave Desert, xiv, 122–23, 126–27
Mondragón, Roberto, 170
monsoons, 10, 38, 68, 72–73, 89–90
Montana, 125
Monte Vista Elementary School
  (Albuquerque), 38
Montiel, Miguel, 186–87
Monument Valley, 60
Moody, Herman, 133–34
Moore, Bob, 102–3, 110–11
Moore, Roger, 131
Moreno, Antonio, 167–68
Morfí, Juan Agustin, 25
Mormons (Latter Day Saints), 74,
  124, 146, 147
Mortgage Investment Company,
  103
Moss, Elisabeth, 90
Motorola, 63
Moulin Rouge casino, 129, 133–36,
  140
Mount Cristo Rey, 98–99, 101,
  112–17

Mount Eolus, 234
Mount Kimball, 92
Mount Lemmon, 68
Mount San Jacinto, xiii
Muench, David, 61
Mulroy, Patricia, 237–41, 251
multiculturalism, 163, 177, 219–20,
    224–31. *See also* racial-ethnic
    politics
Murchison family, 103
"My Dog Barks" (Baca), 176

Naabik'ítáti' Committee, 41
NAACP, 110, 134
Nabhan, Gary, xxiii
Nahuatl language, 9, 69
Nailor, Gerald, 41–42
Namingha, Dan, 69
Nampe Pueblo, 19
Napoles, Cecilia, 106
Naranjo, Domingo, 19
Naranjo Morse, Nora
    Numbe Whageh, 224–27
National Forest Service, 92
National Geographic, 60
National Guard, 168–70
National Hispanic Cultural Center, 6
National Indian Youth Council, 165
National Park Service, 13, 202
National Welfare Rights
    Organization, 138
Native America Calling, 226–27
Native Americans. *See* Indians;
    Puebloans; *and specific peoples; and*
    tribes
Native Seed Search, 223–24
Navajo Dam, 233–34
Navajo (Diné), xv, 5, 11, 17–18, 21,
    23, 26–28, 34, 36–38, 41–42, 190,
    245
Navajo Mountain, 234
Navajo National Monument, 60
Navajo Nation Council, 41
Navajo (train), 35
Neal, Joe, 134
Nebraska, 63, 234

Nellis Air Force Base, 136
Nelson, Edward, 75
Nelson, Pete, 249
Neutra, Richard, 83
Nevada, 61, 234. *See also* Las Vegas;
    and other specific locations
    Blacks and, 132–35
    gambling and, 125–30, 141, 144
    statehood and, 124
    Treaty of Guadalupe Hidalgo
        and, 26
    unions and, 142–44
    water and, 53–54, 237–38, 248
    welfare and, 137–41
Nevada Board of Education, 132
Nevada State Senate, 134
"New Colossus, The" (Lazarus),
    230–231
New Mexico, xiv, xx–xxi, 33–34, 37,
    39–40, 51, 61, 159, 199, 200, 224.
    See also Albuquerque; Sante Fe; *and*
    other specific locations
    Anglos and, 28–29
    Apache and, 73
    AT&SF Railway and, 29–30, 36
    Civil War and, 26–27, 32
    cultures and ethnic politics of, 5–6,
        7–13, 25, 163–64, 224–31, 174
    Spanish conquest and, 13–14
    Spanish land grants and, 164–65
    statehood and, 31–32, 37–39
    Treaty of Guadalupe Hidalgo and,
        25–28, 163
    voting rights and, 32
    water and, 53–54
New Mexico Scorpions, 158
New Mexico State Police, 168
New Mexico Territorial Fair, 37
New Mexico Town Company, 30
New Orleans, 28
New Republic, 183
Newsom, Gavin, 194
Newsweek, 107, 130
Newton, Wayne, 130
New York City, xx–xxi, 77, 79, 103,
    214, 231

New Yorker, 156
New York–New York Hotel and
  Casino, 149
New York State, 146
  Capitol Building, 85
New York Stock Exchange, 147
New York Times, 154, 196, 217,
  238–39
New York Yankees, 59
Nicaraguan migrants, 206
Nile River, 47
Niza, Marcos de, 9
Nogales, Arizona, 75, 206, 191
Nolan, Frederick, 107
No More Deaths / No Más Muertes,
  207
North, Oliver, 92
North American Free Trade
  Agreement (NAFTA), 111
North Korea, 122
Nueva España, 17, 191
Nuevomexicanos, 26, 31–33, 99–100.
  See also Mexican Americans
Nuevo Mexico, Spanish rule and,
  16–25, 99
Nunez-Guardado, Elias, 201–2
Nuwuvi people (Southern Paiute), 124

Obama, Barack, 133
Ochoa, Hector, 203–4
Ohkay Ohwingeh Pueblo, 19
Ojibwe people, 139
Oklahoma, 182
Oñate, Juan de, 14–17, 22, 33, 98–99,
  102, 224–31
O'odham language, 47, 71–72
O'odham peoples, xiv–xv, xxii–xxiii,
  54, 69–71
Operation Gatekeeper, 206
Operation Hold the Line, 206
Operation Lone Star, 214
Oracle Junction, Arizona, 68
Orange County, California, 175
Oregon, 150–51, 223
organized crime, 125–29, 136, 146,
  147

Organ Pipe National Monument,
  202–3
Orleck, Annelise, 136–38
Orme, Lin, 56–57
Orndorff, Bertram, 101
Ortiz, Alfonso, 19
Ortiz, Gaspar, 28
Otermín, Antonio de, 21
Otero, Daniel, 86–87
Otero, Jose Luis, 89
Otero, Lydia, 77–78, 86–90
Our Lady of Fatima, 114
Oyster Bar, 141–42

Pacific Coast, 69
Pacific Islanders, 220
Padilla, Cecelia "Sheila," 159–60,
  171, 174
Painted Desert, 200
Pair O' Dice Club, 125
Pajarito Plateau, xv
Palace of the Governors (Santa Fe),
  17, 21
Palace Station resort, 141–42
Palm Springs, California, xiii, 83
Palo Verde Irrigation District, 244
Panama Canal, 33, 52
Pappageorge, Ted, 143–45
Parada del Sol Parade and Trail's End
  Festival, 65
Paso del Norte Bridge, 210
Pastori, Frank, 131–32
Patriarca crime family, xix, 147
Paxton, Ken, 215
Payson, Arizona, 42
Pecos, Regis, 23
Pecos, New Mexico, 9, 167
Pecos River, 27, 257
Pedro Páramo (Rulfo), 4
Pee Posh people, 245
People, 107
Peralta, Pedro de, 17
Perea, Don José Leandro, 29–30
Pérez Meza, Luis, 75
Permian Sea, 199
Perry, Katy, 151

Peter, George, 184
Petrified Forest, 41, 61
Philippines, 33
Phoenix, xvi–xvii, xxi, 42–47, 49–52, 55–63, 65–67, 69, 111, 126, 128, 165–67, 181–98, 208, 219, 239–43, 245–46, 250–52, 258. *See also* Salt River Valley *and* Valley of the Sun
Phoenix chamber of commerce, 50–51, 59, 61–63
Phoenix Country Club, 62–63, 182
Phoenix Mountains, 189–90
Phoenix New Times, 62–63
Phoenix Open golf tournament, 62–63
Phoenix police, 184
Phoenix Sky Harbor, 45
Phoenix Water Service, 240
Piedra River, 234
Piestewa Peak, 193
Pilgrims, 17
Pima County, Arizona, 90
Piro people, 8, 11, 99
Pitbull, 152
Pizarro, Francisco, 33
Playboy Club (Chicago), 130
Pojoaque Pueblo, 230
Polk, James, 28
Pony Express Week, 65
Pony saloon, 100
Po'pay (Tewa leader, also Popé), 19–21
Po-sogeh Pueblo, 19
Poster, Corky, 81–82
poverty, 102–3, 187–88
Powell, Bryant, 243, 246
Pratt, Richard Henry, 161
Prendergast, Curt, 207
Prescott, Arizona, 74
Presley, Elvis, 131
Primitive Indians of the Painted Desert (film), 36
"Proclamation" (Zepeda), 72
Programa de Industrialización Fronteriza, 103
Prohibition, 101, 126

ProPublica, 141, 217
Puebloans, xv, 7–25, 32, 34–36, 48, 124, 160, 224–25, 256. *See also* Hopi; *and specific groups*
Pueblo Bonito, 7
Pueblo Center Redevelopment Project, 76–78, 167
Pueblo Revival style, 184, 191–92
Pueblo Revolt (1680), 5, 18–23, 99, 105
Puerto Ricans, 77
Puerto Rico, 33

Quetzalcoatl, 256
Quiñones, Ramon, 206
Quitman Mountains, 96

racial covenants, 184–86
racial-ethnic politics, 163–64, 166, 182–88, 228
Radium Springs, New Mexico, 109
Ragsdale, Eleanor Dickey, 182–85, 195
Ragsdale, Hartwell, 182
Ragsdale, Lincoln, 182–85, 195
Ramada House, 81–82, 90–91
Rams, Dieter, 199
Reagan, Ronald, 137, 204, 218
real estate development, 89–90, 147–49, 158, 184–98, 235–38, 242–43, 250–53, 258–59
Reconquista (1692), 22–23
Reconstruction, 32
Red Cross, 211
redlining, 76, 182
Red Nation, 229
Red Power movement, 165
Remain in Mexico policy, 212, 216
Reno, Nevada, 124
Reynolds, Larry T., xviii–xix
Richardson, Bill, 5, 176
Richardson, H. H., 85
Rieveschl, George, 92
Rieveschl, Rose, 92
Rieveschl House, 92–93, 252
Rillito River, 91

Rio Arriba County, New Mexico, 228–29
Rio Chama, 15
Rio Grande, xiv, 4, 6, 8, 14–15, 29, 38, 44, 96–100, 157–58, 167, 208–9, 218–19, 234, 258
Rio Grande Gorge, 218
Rio Grande Rectification Project, 218–19
Rio Grande Valley, 3, 10, 24–25, 40
Rio Hotel and Casino, 153
Rittgers, Mateele, 110
Rivera, Carlos, 201–3
Rivera, Geraldo, 105–6, 108, 110
Rivera, Reynaldo "Sonny," 225
    La Jornada (with Sabo), 227–28
    Oñate statue, 225–26, 228–29
River Mountains Water Treatment Facility, 240
Riviera casino, 129
Robb, Inez, 61–62
Robles, Jesús, 190–92
Robles, Maria de la Cruz "Chita," 86–88
Robles, Pa' Luis, 88
Rocky Mountains, xv, 7, 25, 36, 53, 99
Roden Crater, 85
Rogers, Jonathan, 103–4, 111
Roosevelt, Eleanor, 36
Roosevelt, Franklin Delano, 112
Roosevelt, Theodore, 52
Roosevelt Dam, 52, 55, 56
Roosevelt Irrigation District, 56
Roosevelt Reservoir, 57
Rose, Charlie, 146
Rosenstein, Simon, 28–29
Royal Nevada casino, 129
Ruby Canyon, 53
Rudolph, Paul, 79
Ryan, Thomas Fortune, 108
Ryan, Thomas Fortune, III, 108–10

Saarinen, Eero, 79
Sabo, Betty, 225–27
    La Jornada (with Rivera), 227–28
Sacramento Mountains, 199–200

Sacred Heart Church (Phoenix), 186
Saddle Island, 237
Sahagún, Louis, 204
Sahara casino, 130, 137
Salazar y Villaseñor, José Chacón Medina, 158
Salomon, Jane London, 82
Salomon, Peter, 82, 90
Salt Lake City, 125, 147
Salton Sea, 244
Salt River, xiv, 43–44, 47–52, 57, 69, 182–84, 246
Salt River Project (formerly Salt River Valley Water Users' Association), 52, 56, 65, 193, 195
Salt River Valley, xv, 42, 44, 50–51, 55, 62, 194, 256. See also Phoenix and Valley of the Sun
Salt War of San Elizario, 95–96
Salvadoran migrants, 201–6, 212, 216–17
San Andres Mountains, 199
San Antonio, Texas, 25, 95
Sanctuary Movement, 205–6
Sandia Mountains, xv, xx, 3, 10–11, 29, 38, 42
Sandia National Laboratory, 111
Sandia Pueblo, 5, 13, 20, 23–24, 158
San Diego, xiii, 60, 74, 206, 211, 244, 246, 248, 250
Sandinistas, 206
Sands casino, 136, 139, 144, 146
San Felipe de Neri church, 24, 29
San Fernando Valley, 243
San Francisco, 59, 65
San Francisco Peaks, 222
San Gabriel, Nuevo Mexico, 15–17
Sangre de Cristo Mountains, 17, 25, 53
San Ildefonso Pueblo, 22, 229
San José de Cristo Rey, 112
San Juan Mountains, 233–34
San Juan River, 233–34
San Luis, Mexico, 201–2
San Luis Potosí, xiv
San Marcos Pueblo, 19

San Rafael Valley, 191
San Salvador, 201
Santa Bárbara, Mexico, 14
Santa Catalina Mountains, xv, 68, 71, 81, 83–85, 90, 92
Santa Clara Pueblo, 225
Santa Cruz River, xxiii, 47, 48, 69, 73, 86–89
Santa Cruz Valley, 88
Santa Fe, xvii–xviii, 3, 17–22, 25, 28–29, 35–36, 40, 53–55, 99, 125, 158–59, 161, 176, 190–91, 229
Santa Fe New Mexican, 23
Santa Fe Ring, 32, 164
Santa Fe River, 21
Santa Monica, 126
Santa Monica Mountains, 93
Santander Province, Spain, 10
Sante Fe Bridge (El Paso), 101
Santillanes, Mille, 225–26
San Xavier del Bac Mission, 73–74, 223
San Xavier Reservation, 224
San Ysidro port of entry, 211
Sarno, Jay, 153
Saslow, Eli, 196
Saudi Arabia, 92
Scottsdale, Arizona, xvii, 43, 64–66, 188
Scottsdale Public Art Program, 65
Seal, Mark, 148
Sea of Cortez (Gulf of California), xxii, 53, 86, 200, 249
Second Great Migration, 134
Section 8 housing, 196
Sedway, Moe, 126, 128
segregation, 74–76, 88–89, 135–36, 165, 182–85, 205, 220
Sellers, Colonel D. K. B., 37–38
Sells (Tohono O'odham reservation), 72
Sen, Sreshtha, 165–66
Sentinel Mountain, xxii
Sentinel Peak (A Mountain), 86
Seoul, South Korea, 122
September 11, 2001, attacks, 5

Sepulveda, Robert, 89
Sepulveda, Shina, 196
Seri people, 86
Seven Cities of Cíbola, 9
Shakira, 151
Shapiro, Jack, 146
Sheep Range, 133
Siegel, Benjamin "Bugsy," 125–28, 147–48, 153, 183
Sierra Blanca, Texas, 96
Sierra Madres, 14, 99
Sierra Nevada, xv, 124
Silko, Leslie Marmon, 235
Silver City, New Mexico, 96
"Silver Water Tower" (Baca), 160–62
Simon Canyon, 233
Sinaloa Cartel, 210
Sinatra, Frank, 131, 135–36, 148
slavery, 18, 26, 31, 163, 227, 228
  See also Genízaros
Smeltertown, Texas, 112–13, 116
Smith, John Y. T., 48–49
Smythe, William, xvi, 33–34, 50–51, 218
Soler, Urbici, 113–16
  Christ of the Rockies, 116–17
Sonoran Desert, xiv, xxii–xxiii, 7, 44, 47–48, 69–70, 72, 81, 86, 95, 188, 251, 252
Sonoyta, Mexico, 202
Sorensen, Kathryn, 192–95
Soul Brothers Motorcycle Club, 140
Southern Nevada Water Authority (SNWA), 237–41
Southern Pacific Railroad, 113
South Jordan, Utah, 246
South Korea, 132
Southside Presbyterian Church (Tucson), 204–7
Spain, 163
Spanish-American War (1898), 33
Spanish land grants, 86, 157–58, 164–66, 169
Spanish language, 160
Spanish rule, xv–xvi, xxii, 7–25, 41, 73, 98, 124, 163–64, 184, 224–31

Sperry Rand, 63
Stacey, Joe, 64
Stanford, Rawghlie C., 58
Stanford University, 57
St. Anthony's Catholic Orphanage
    (Albuquerque), 160–62
Stardust casino, 129–31
State Department, 237
Statue of Liberty, 230–31
Stevens, Derek, 153–55
Stevens, Greg, 153–54
Stieve, Robert, 45–46, 64
Stillwater, Oklahoma, 165
St. Louis Blues, 152
Storch, Kirk, 76
Streeter, David, 83–84
Strong Brothers general store, 30
Stupak, Bob, 153
Suazo, Jerry, 161
Sullivan, Ed, 130
Sullivan, Theresa, 80
Sumerians, 47
"Sunbelt Syndrome, The" (Hertz,
    Haywoode, and Reynolds), xviii–xx,
    xxiv, 156
Sun City, Arizona, 61
Sunset Station Casino, 235
Super Bowl, 153–54
Super Chief (train), 35
Superfund sites, 186
Superstition Mountains, 61, 67, 242
Superstition Vistas report, 243–44
sustainability, 192–95, 259
Sutter's Mill, 28
Sweeney, John, 142–43
Swilling, Tragic Jack, 49
Syrian migrants, 75

Taliesin West museum, 188–90
Tallulah, Mississippi, 134
Tampa, Florida, 60
Tanoan language, 10
Tano people, 21
Taos, New Mexico, 15–16, 35, 167,
    218
Taos Pueblo, 161

Tapachula, Mexico, 201
Tapia, Jose, 89
Tarahumara people, xiv
Tartesso, development Arizona,
    251–53
Tempe, Arizona, 51, 62, 193, 222
Temple Emanu-El (Tucson), 205–6
Temporary Protected Status (TPS),
    206, 214–15
Teotihuacan, 256
Tewa people, xv, 8, 11, 19, 225–26
Texas, xvii, xviii, 5, 28, 73, 95, 101,
    159, 200, 205, 214
Texas A&M, 193
Texas Monthly, 104, 108
Texas Rangers, 96
Texas Republic, 25, 99
Texas Tech, 190
Thai Americans, 220
Thomas, E. Parry, 146–47
Thompson, Hunter S., 150
Three Rivers Ranch (New Mexico),
    108
Three Sisters volcanoes, 4
Thunderbird Hotel, 121–22, 129,
    131, 143, 147
Thunderbirds, 62–63
Tierra Amarilla land grant, 164
Tigris and Euphrates Rivers, 47
Tiguex province, 10–13, 16, 23–24,
    173
Tijeras Pass, 3–4
Tijerina, Reies López, 164–66
Tijuana, 103
Time, 104
Title 42, 212–13, 216
Tiwa people, 10–13
Tohono O'odham Grammar, A
    (Zepeda), 71–73
Tohono O'odham language, 71–72
Tohono O'odham Legislative
    Council, 72
Tohono O'odham people, xxii, xxiii,
    69–73, 224
Tohono O'odham reservation, 72
Tonopah, Nevada, 125

Tony Lama Boots company, 103
Torres, Alva, 78
tourism, 35–40, 60–66, 78, 101–2, 123, 129, 155
Touton, Camille, 247
Toy, Dorothy, 122
Treasure Island Hotel, 148–49
Treviño, Juan Francisco, 19
Tricklock theater company, 6
Trujillo, Joseph de, 20
Trujillo, Miguel, 32
Trujillo, Simón Ventura, 163–64
Truman, Harry, 190
Trumka, Rich, 144
Trump, Donald, 211–12, 215–16
Tséghahoodzání (Window Rock), 41
Tsoodził (Mount Taylor), 27
Tsyitee, Ed, 161
Tubac land grant, 86
Tucson, xvi, xxi–xxiv, 67–93, 133, 190, 204–8, 219, 221–24, 239, 245
Tucson Basin, 48, 223
Tucson Community Center, 76
Tucson Daily Citizen, 75, 77–78, 88, 204–5
Tucson Historic Preservation Foundation, 85
Tucson Mountains, 82, 93
Tucson Water Department, 74
Tucson Weekly, 76, 89
Tulsa race riot (1921), 182
Turgenev, Ivan, 177
Turkish migrants, 214, 217
Turrell, James
  The Light Inside, 85
  Roden Crater installation, 85
Tuskegee Airmen, 182

UNICEF, 212
UNIDAD, 77
Union Army, 32
Union Pacific Railroad, 74
United Mine Workers of America, 144
U.S. Air Force, 182–83, 243
U.S. Army, 26, 29, 50, 59, 103, 132

U.S. Cavalry, 37, 258
U.S. Citizen Identification Cards, 86
U.S. Congress, 28, 32, 37, 53, 55, 138, 206, 245
U.S. Marines, 32
U.S. Marshals, 74
U.S. Senate, 31
U.S. Supreme Court, 32
University of Arizona (U of A), xxiii, 57, 69, 71–72, 74, 78, 86, 111, 133, 222
University of Miami, 175
University of Nevada, Las Vegas (UNLV), 123, 138, 141, 144–45, 237
  Marjorie Barrick Museum of Art, 133
University of New Mexico (UNM), 4, 5, 38, 123, 165–70 228
University of Texas at El Paso (UTEP), 102, 111
Ur, 47
urban renewal, 76–78, 89–90, 220
Usery Mountains, 48
USO, 122
Utah, 26, 53, 54, 61, 127, 146
Utah Territory, 124
Ute people, 15, 23, 25, 54

Valenzuela, Enrique, 210–12, 218
Valley National Bank, 59, 127–28, 183, 185
Valley of the Sun, 43–44, 61–63, 65, 185, 187, 189–90, 196, 242–44, 249–52. *See also* Phoenix *and* Salt River Valley
Vargas, Diego de, 22–23, 99
Vegas Golden Knights, 221, 240
Vegas Vickie, 152–53
Venezuelan migrants, 214–15, 217–18
Venturi, Robert, xviii, 122
Verde River, 43, 48, 56, 58, 246
Vietnam War, 169, 176
Viewpoint House, 82–84, 191
Vigil, Francisco Montes, 157–58

Villa de Alburquerque settlement, 24, 26, 28–30, 39, 225. *See also* Albuquerque
Village Baker of Flagstaff, 222–23
Villegas (Spanish soldier), 12
Virginia City, Nevada, 124
Vivero (Spanish soldier), 16

Walk the Moon, 152
Wallace, Norman G., 61
Wall Street Journal, 149
Waltz, Jacob, 67–68
Wampanoag language, 71
Warlpiri language, 71
Washington, Booker T., 182
Washington, Dinah, 135
Washington, Johnnie, 110
Washington, D.C., 214, 242
Washington Post, 102, 143
water, xiv–xv, xxii, 57, 88, 97, 186, 188, 192–95, 236–51, 255–59. *See also* drought
"Water in the Magic Land" (Carlson), 58
Wayson, D. Boone, 149–50
Webb, Del, 59, 61, 63, 127, 183
Weizenbaum, Joseph, 205
Wesley, Mary, 139
West, Charles, 136, 138
Western Area Power Administration, 234
West Texas, xiii, 200
wheat, 48–50
Whitaker, Matthew, 182
White Elephant saloon, 30–31
White Sands National Park, 199
white supremacy, 31, 163, 188, 227
White Tank Mountains, 250
Why, Arizona, 203
Wickenburg, Arizona, 48–49
Wild Card Weekend, 153

Wilder, Joseph, 88
Wiley, George, 138–39
Wilkinson, Gerald, 165
Wills, Laura, 83
Wilson, Chip, 243, 246
Wilson Mountain (Colorado), 234
Window Rock, Arizona, 41
Wing, Paul, 122
Wolfe, Tom, 122
Wong, Brian, 224
Wordsworth, William, 177
World Series, 153
World War II, 3–4, 32, 39, 58–61, 87, 103, 122, 134, 182, 236
Wrangler company, 111
Wright, Frank Lloyd, 46, 64, 188–90
Wynn, Steve, 146–50, 153, 155, 238–39
Wyoming, 53–54

Xavier, San Francisco, 24
Xenome, Diego, 19

Yale School of Architecture, 79
Yellow River, 47
Yi Nanyŏng (mother of Sue Kim), 121–22, 130–31
Young, Brigham, 124, 147, 193
Ysleta, Texas, 105
Yuma, 48, 54
Yunge Oweenge Pueblo (*later* San Gabriel), 15–16
Yu the Great, 47

Zacatecas silver mines, 14
Zaldívar, Juan de, 16
Zepeda, Ofelia, 69–73, 93
Zerilli, Anthony, 146
Zia Pueblo, 157
Zona Libre, 103
Zuni people, 7–9, 23, 256

A NOTE ABOUT THE AUTHOR

Kyle Paoletta's reporting and criticism has appeared in *Harper's*, *New York*, *The Nation*, *The New Republic*, *The New York Times*, *The Believer*, *Columbia Journalism Review*, *The Baffler*, *High Country News*, and *Boston*. Kyle holds an MFA in fiction from Columbia University and previously worked at *GQ* and *New York*. He grew up in Albuquerque, New Mexico, and lives in Cambridge, Massachusetts. This is his first book.

A NOTE ON THE TYPE

This book was set in Janson, a typeface long thought to have been made by the Dutchman Anton Janson, who was a practicing typefounder in Leipzig during the years 1668–1687. However, it has been conclusively demonstrated that these types are actually the work of Nicholas Kis (1650–1702), a Hungarian, who most probably learned his trade from the master Dutch typefounder Dirk Voskens. The type is an excellent example of the influential and sturdy Dutch types that prevailed in England up to the time William Caslon (1692–1766) developed his own incomparable designs from them.

Composed by North Market Street Graphics,
Lancaster, Pennsylvania

Printed and bound by Berryville Graphics,
Berryville, Virginia

Designed by Soonyoung Kwon